Encyclopedia of Earthquake Research and Analysis: Analytical Studies

Volume II

Encyclopedia of Earthquake Research and Analysis: Analytical Studies
Volume II

Edited by **Daniel Galea**

New York

Published by Callisto Reference,
106 Park Avenue, Suite 200,
New York, NY 10016, USA
www.callistoreference.com

**Encyclopedia of Earthquake Research and
Analysis: Analytical Studies
Volume II**
Edited by Daniel Galea

International Standard Book Number: 978-1-63239-235-0 (Hardback)

Contents

Preface

To minimize human and material losses due to inevitable occurrence of earthquakes, their study and understanding is of utmost importance. This book discusses different aspects related to earthquake research and its analysis. Some of the many topics included in this book are statistical seismology studies, the latest techniques and advances on earthquake precursors and forecasting, and also, new ways for immediate detection, data acquisition and its interpretation. The book covers a wide range of topics, from theoretical advances to practical applications.

This book is the end result of constructive efforts and intensive research done by experts in this field. The aim of this book is to enlighten the readers with recent information in this area of research. The information provided in this profound book would serve as a valuable reference to students and researchers in this field.

At the end, I would like to thank all the authors for devoting their precious time and providing their valuable contribution to this book. I would also like to express my gratitude to my fellow colleagues who encouraged me throughout the process.

Editor

Part 1

Statistical Seismology

The Weibull – Log-Weibull Transition of Interoccurrence Time of Earthquakes

Tomohiro Hasumi[1], Chien-chih Chen[2], Takuma Akimoto[3] and Yoji Aizawa[4]
[1]Division of Environment, Natural Resources and Energy,
Mizuho Information and Research Institute, Inc., Tokyo
[2]Department of Earth Sciences and Graduate Institute of Geophysics,
National Central University, Jhongli, Taoyuan
[3]Department of Mechanical Engineering, Keio University, Yokohama
[4]Department of Applied Physics, Advanced School of Science and Engineering,
Waseda University, Tokyo
[1,3,4]Japan
[2]Taiwan

1. Introduction

Earthquakes are great complex phenomenon characterized by several empirical statistical laws (1). One of the most important statistical law is the Gutenberg - Richter law (2), where the cumulative number of $n(> m)$ of magnitude m satisfy the following relation:

$$\log n(> m) = a - bm, \tag{1}$$

where a and b are constants. b is so-called b-value and is similar to unity. Another important statistical law is a power law decay of the occurrence of aftershocks, called Omori law (3).

The time intervals between successive earthquakes can be classified into two types: interoccurrence times and recurrence times (4). Interoccurrence times are the interval times between earthquakes on all faults in a region, and recurrence times are the time intervals between earthquakes in a single fault or fault segment. For seismology, recurrence times mean the interval times of characteristic earthquakes that occur quasi-periodically in a single fault. Recently, a unified scaling law of interoccurrence times was found using the Southern California earthquake catalogue (5) and worldwide earthquake catalogues (6). In Corral's paper (6), the probability distribution of interoccurrence time, $P(\tau)$, can be written as

$$P(\tau) = Rf(R\tau), \tag{2}$$

$$f(R\tau) = C\frac{1}{(R\tau)^{1-\gamma}} \exp(-(R\tau)^{\delta}/B), \tag{3}$$

where R is the seismicity rate. He has found that $f(R\tau)$ follows the generalized gamma distribution. In equation (3), C is a normalized constant and is $C = 0.50 \pm 0.05$. γ, δ, and B are parameters estimated to be $\gamma = 0.67 \pm 0.05$, $\delta = 0.98 \pm 0.05$, and $B = 1.58 \pm 0.15$. It should

Catalog Name	Coverage	Term	Number of Earthquakes	m_{min}	m_c^0
JMA	25° –50° N and 125° –150° E	01/01/2001–31/1/2010	170,801	2.0	2.0
SCEDC	32° –37° N and 114° –122° W	01/01/2001–28/2/2010	116,089	0.0	1.4
TCWB	21° N–26° N and 119° –123° E	01/01/2001–28/2/2010	189,980	0.0	1.9

Table 1. Information on earthquake catalogues.

be noted that the interoccurrence times were analyzed for the events with the magnitude m above a certain threshold m_c under the following two assumptions: (a) earthquakes can be considered as a point process in space and time; (b) there is no distinction between foreshocks, mainshocks, and aftershocks. It has been shown that the distribution of the interoccurrence time is also obtained by analyzing aftershock data (7) and is derived approximately from a theoretical framework proposed by Saichev and Sornette (8; 9). Abe and Suzuki showed that the distribution of the interoccurrence time, $P(> \tau)$, can be described by q-exponential distribution with $q > 1$, corresponding to a power law distribution (10), namely,

$$P(> \tau') = \frac{1}{(1 + \epsilon\tau')^\gamma} = e_q(-\tau'/\tau_0) = [(1 + (1 - q)(-\tau'/\tau_0))^{\frac{1}{1-q}}]_+, \qquad (4)$$

where $q_t, \tau_0, \gamma,$ and ϵ are positive constants and $([a]_+ \equiv \max[0, a])$.

It has been reported that the sequence of aftershocks and successive independent earthquakes is a Poisson process (11; 12). However, recent works show that interoccurrence times are not independent random variables, but have "long-term memory" (13–16). Since an interoccurrence time depends on the past, it is difficult to determine the distribution of interoccurrence times theoretically. Therefore, the determination of the distribution of interoccurrence times is still an open problem. Moreover, an effect of a threshold of magnitude on the interoccurrence time statistics is unknown. We study the distribution of interoccurrence times by changing the threshold of magnitude. In this chapter, we review our previous studies (17; 18) and clarify the Weibull - log-Weibull transition and its implication by reanalyzing the latest earthquake catalogues, JMA catalogue (19), SCEDC catalogue (20), and TCWB catalogue (21). This study focuses on the interoccurrence time statistics for middle or big mainshocks.

2. Data and methodology

2.1 Earthquake catalogue

To study the interoccurrence time statistics, we analyzed three natural earthquake catalogues of the Japan Metrological Agency (JMA) (19), the Southern California Earthquake Data Center (SCEDC) (20) and the Taiwan Central Weather Bureau (TCWB) (21). Information on each catalogue is listed in Table 1, where m_{min} corresponds to the minimum magnitude in the catalogue and m_c^0 is the magnitude of completeness, that is the lowest magnitude at which the Gutenberg - Richter law holds. We basically consider events with magnitude greater than and equal to m_c^0 because events whose magnitudes are smaller than m_c^0 are supposedly incomplete for recording.

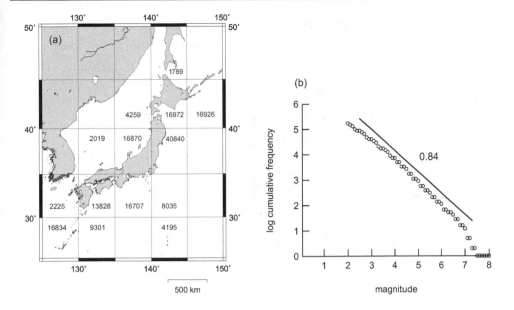

Fig. 1. Information on the Japan Metrological Agency (JMA) earthquake catalogues. (a) covering region. The number of each cell means the number of earthquakes. (b) the magnitude distribution. $b = 0.84$ is calculated from the slope of the distribution.

2.1.1 Japan Metrological Agency (JMA) earthquake catalogue

JMA catalogue is maintained by the Japan Metrological Agency, covering from 25° to 50° N for latitude, and from 25° to 150° E for longitude [see Figure 1 (a)] during from 1923 to latest. This catalogue consists of an occurrence of times, a hypocenter, a depth, and a magnitude. In this chapter, we use the data from 1st January 2001 to 31st January 2010. As can be seen from Figure 1 (b), the distribution of magnitude obeys the Gutenberg - Richter law and m_c^0 is estimated to be 2.0.

2.1.2 Southern California Earthquake Data Center (SCEDC) earthquake catalogue

SCEDC catalogue is maintained by the Southern California Earthquake Data Center, covering from 32° to 37° N for latitude, and from 114° to 122° W for longitude [see Figure 2 (a)] during from 1932 to latest. The information of an earthquake, such as an occurrence of times, a hypocenter, a depth, and a magnitude, is listed. Here, we analyze the earthquake data from 1st January 2001 to 28th February 2010. In Figure 2 (b), we demonstrate the magnitude distribution, and we obtain $b = 0.97$.

2.1.3 Taiwan Central Weather Bureau (TCWB) earthquake catalogue

TCWB catalogue is maintained by the Central Weather Bureau, covering from 21° to 26° N for latitude, and from 119° to 123° E for longitude [see Figure 3 (a)]. This catalogue consists of an occurrence of times, a hypocenter, a depth, and a magnitude. We use the data from 1st January 2001 to 28th February 2010. As shown in Figure 3 (b), the Gutenberg - Richter law is valid in a

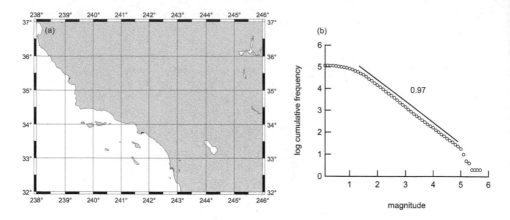

Fig. 2. Southern California Earthquake Data Center (SCEDC) earthquake catalogue information. (a) covering region. (b) the magnitude distribution. $b = 0.97$ is calculated from the slope of the distribution.

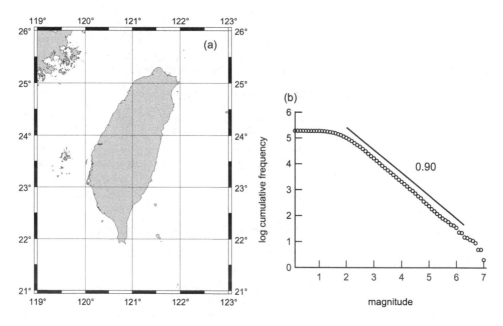

Fig. 3. Information on the Taiwan Central Weather Bureau (TCWB) earthquake catalogue. (a) covering region. (b) the magnitude distribution. $b = 0.90$ is calculated from the slope of the distribution.

magnitude range, $1.9 \leq m \leq 6.7$. b-value is calculated from the slope of the distribution, and is estimated to be 0.90.

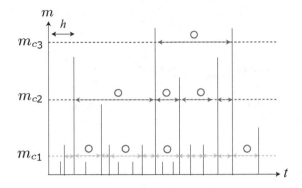

Fig. 4. Schematic diagram of the interoccurrence time of our analysis for different threshold of magnitude m_c. Circles (○) satisfy the condition. We analyze interoccurrence times greater than or equals to h.

2.2 Methodology (How to detect the appropriate distributions)

Our method is similar to the that of previous works (17; 18) (see Figure 4).

1. We divided the spatial areas into a window of L degrees in longitude and L degrees in latitude.
2. For each bin, earthquakes with magnitude m above a certain cutoff magnitude m_c were considered.
3. We analyzed interoccurrence times greater than and equals to h day.

For each bin, we analyzed interoccurrence times using at least 100 events to avoid statistical errors. h and L are taken to be 0.5 and 5, respectively. It is noted that for SCEDC and TCWB, we analyze earthquake covering the whole region. As shown in Figure 4, we investigated the interoccurrence time statistics for different 16 regions (14 regions in Japan, Southern California, and Taiwan). Aftershocks might be excluded from the study based on the information from previous studies (6; 12).

One of our main goals in this chapter is to determine the distribution function of the interoccurrence time. Here, we will focus our attention on the applicability of the Weibull distribution P_w, the log-Weibull distribution P_{lw} (22), the power law distribution P_{pow} (10), the gamma distribution P_{gam} (in the case of $\delta = 1$ in the paper (6)), and the log normal distribution P_{ln} (23), which are defined as

$$P_w(\tau) = \left(\frac{\tau}{\beta_1}\right)^{\alpha_1 - 1} \frac{\alpha_1}{\beta_1} \exp\left[-\left(\frac{\tau}{\beta_1}\right)^{\alpha_1}\right], \tag{5}$$

$$P_{lw}(\tau) = \frac{(\log(\tau/h))^{\alpha_2 - 1}}{(\log \beta_2)^{\alpha_2}} \frac{\alpha_2}{\tau} \exp\left[-\left(\frac{\log(\tau/h)}{\log \beta_2}\right)^{\alpha_2}\right], \tag{6}$$

$$P_{pow}(\tau) = \frac{1}{(1 + \beta_3 \tau)^{\alpha_3}}, \tag{7}$$

$$P_{gam}(\tau) = \tau^{\alpha_4 - 1} \frac{\exp(-\tau/\beta_4)}{\Gamma(\alpha_4)\beta_4^{\alpha_4}}, \tag{8}$$

$$P_{ln}(\tau) = \frac{1}{\tau\beta_5\sqrt{2\pi}} \exp\left[-\frac{(\ln(\tau) - \alpha_5)^2}{2\beta_5^2}\right], \tag{9}$$

where α_i, β_i, and h are constants and characterize the distribution. $\Gamma(x)$ is the gamma function. i stands for an index number; $i = 1, 2, 3, 4$, and 5 correspond to the Weibull distribution, the log-Weibull distribution, the power law distribution, the gamma distribution, and the log normal distribution, respectively.

The Weibull distribution is well known as a description of the distribution of failure-occurrence times (24). In seismology, the distribution of ultimate strain (25), the recurrence time distribution (26; 27), and the damage mechanics of rocks (28) show the Weibull distribution. In numerical studies, the recurrence time distribution in the 1D (4) and 2D (29) spring-block model, and in the "Virtual California model" (30) also exhibit the Weibull distribution. For $\alpha_1 = 1$ and $\alpha_1 < 1$, the tail of the Weibull distribution is equivalent to the exponential distribution and the stretched exponential distribution, respectively. The log-Weibull distribution is constructed by a logarithmic modification of the cumulative distribution of the Weibull distribution. In general, the tail of the log-Weibull distribution is much longer than that of the Weibull distribution. As for $\alpha_2 = 1$, the log-Weibull distribution is equal to a power law distribution. It has been shown that the log-Weibull distribution can be derived from the chain-reaction model proposed by Huillet and Raynaud (22).

To determine the best fitting for the distribution of the interoccurrence time data, we used the root mean square (rms) and Kolomogorov-Smirnov (KS) tests as the measure of goodness-of-fit. The definition of the rms value is

$$\text{rms} = \sqrt{\frac{\sum_{i=1}^{n}(x_i - x_i')^2}{n - k}}, \tag{10}$$

where x_i is actual data and x_i' is estimated data obtained from $P(\tau)$. n and k indicate the numbers of the data points and of the fitting parameters, respectively. In this study, the rms value is calculated using the cumulative distribution for decreasing the fluctuation of the data. The most appropriate distribution is, by definition, the smallest rms value. Also, in order to use the KS test, we define the maximum deviation of static DKS, which is so-called Kolomogorov-Smirnov statistic, as

$$\text{DKS} = \max_i |y_i - y_i'|, \tag{11}$$

where y_i and y_i' mean the actual data of the cumulative distribution and the data estimated from the fitting distribution, respectively. Then, the significance level of probability of the goodness-of-fit, Q, is defined as

$$Q = 2\sum_{i=1}^{\infty}(-1)^{i-1}e^{2i^2\lambda^2}, \tag{12}$$

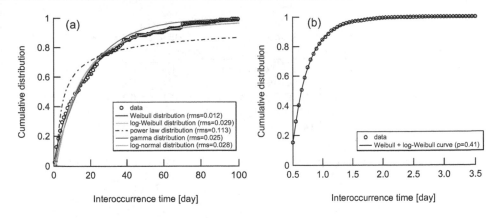

Fig. 5. Cumulative distribution of interoccurrence time at Okinawa region different m_c. The cumulative distribution is plotted by circles. (a) Several fitting curves are represented by lines ($m_c = 4.5$). (b) The superposition of the Weibull and the log-Weibull distribution is represented by line ($m_c = 2.0$).

where

$$\lambda = DKS \left(\sqrt{n'} + 0.12 + \frac{0.11}{\sqrt{n'}} \right),$$ (13)

where n' stands for the number of data points.

It is known that the preferred distribution shows the smallest value of DKS and the largest value of Q (31).

3. Results

3.1 Interoccurrence time statistics in Japan

First, we analyze the JMA data. Here, we consider the two region; Okinawa region ($125°$–$130°$E and $25°$–$30°$N), and Chuetsu region ($135°$–$140°$E and $35°$–$40°$N). The total number of earthquakes in Okinawa and Chuetsu are 16,834 and 16,870, respectively.

The cumulative distributions of the interoccurrence times for different m_c in Okinawa region and in Chuetsu region are displayed in Figures 5 and 6, respectively. We carried out two statistical tests, the rms and the KS test so as to determine the distribution function. The results for large magnitude ($m_c = 4.5$) in Okinawa and Chuetsu are shown in Table 2 and 3, respectively. For Okinawa, we found that the most suitable distribution is the Weibull distribution in all tests. In general, there is a possibility that the preferred distribution is not unique but depends on the test we use. However, the results obtained in Table 2 provide evidence that the Weibull distribution is the most appropriate distribution. As for Chuetsu, by two tests, the preferred distribution is suited to be the Weibull distribution as shown in Table 3, where the Weibull distribution is the most prominent distribution in the two tests. It follows that the Weibull distribution is preferred.

m_c		Distribution X		RMS test	KS test	
Region	distribution X	α_i	β_i [day]	rms [$\times 10^{-3}$]	DKS	Q
	P_w $(i=1)$	0.82 ± 0.007	19.0 ± 0.13	12	0.03	1
	P_{lw} $(i=2)$	3.08 ± 0.06	35.3 ± 0.57	29	0.1	0.15
4.5	P_{pow} $(i=3)$	1.48 ± 0.02	1.04 ± 0.12	113	0.23	3.8×10^{-3}
Okinawa	P_{gam} $(i=4)$	0.96 ± 0.005	19.5 ± 0.24	24	0.07	0.55
	P_{ln} $(i=5)$	2.45 ± 0.02	1.20 ± 0.02	28	0.09	0.33

Table 2. Results of rms value, DKS, and Q for different distribution functions for Okinawa area ($m_c = 4.5$). The error bars mean the 95% confidence level of fit.

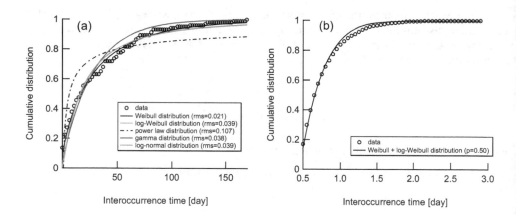

Fig. 6. Cumulative distribution of interoccurrence time for Chuetsu area different m_c. The cumulative distribution is plotted by circles. (a) Several fitting curves are represented by lines ($m_c = 4.5$). (b) The superposition of the Weibull and the log-Weibull distribution is represented by line ($m_c = 2.0$).

m_c		Distribution X		RMS test	KS test	
Region	distribution X	α_i	β_i [day]	rms [$\times 10^{-3}$]	DKS	Q
	P_l $(i=1)$	0.75 ± 0.01	27.6 ± 0.40	21	0.06	0.94
	P_{lw} $(i=2)$	3.12 ± 0.10	51.4 ± 1.39	39	0.14	0.06
4.5	P_{pow} $(i=3)$	1.47 ± 0.03	1.51 ± 0.21	107	0.19	4.8×10^{-3}
Chuetsu	P_{gam} $(i=4)$	0.94 ± 0.009	28.9 ± 0.67	38	0.11	0.27
	P_{ln} $(i=5)$	2.78 ± 0.03	1.33 ± 0.04	39	0.12	0.15

Table 3. Results of rms value, DKS, and Q for different distribution functions for Chuetsu area ($m_c = 4.5$). The error bars mean the 95% confidence level of fit.

m_c Region	distribution X	Weibull distribution α_1	β_1 [day]	Distribution X α_i	β_i [day]	p	RMS test rms [$\times 10^{-3}$]	KS test DKS	Q
4.5 Okinawa	P_{lw} ($i=2$)	0.82 ± 0.007	19.0 ± 0.13	−	−	1	12	0.03	1
	P_{pow} ($i=3$)	0.82 ± 0.007	19.0 ± 0.13	−	−	1	12	0.03	1
	P_{gam} ($i=4$)	0.82 ± 0.007	19.0 ± 0.13	−	−	1	12	0.03	1
	P_{ln} ($i=5$)	0.82 ± 0.007	19.0 ± 0.13	−	−	1	12	0.03	1
4.0 Okinawa	P_{lw} ($i=2$)	0.93 ± 0.009	8.47 ± 0.06	−	−	1	13	0.03	1
	P_{pow} ($i=3$)	0.93 ± 0.009	8.47 ± 0.06	−	−	1	13	0.03	1
	P_{gam} ($i=4$)	0.93 ± 0.009	8.47 ± 0.06	−	−	1	13	0.03	1
	P_{ln} ($i=5$)	0.93 ± 0.009	8.47 ± 0.06	−	−	1	13	0.03	1
3.5 Okinawa	P_{lw} ($i=2$)	1.07 ± 0.008	3.45 ± 0.02	2.07 ± 0.04	6.25 ± 0.09	0.77 ± 0.02	5.3	0.02	1
	P_{pow} ($i=3$)	1.07 ± 0.008	3.45 ± 0.02	1.81 ± 0.04	0.64 ± 0.04	0.94 ± 0.009	7.3	0.03	1
	P_{gam} ($i=4$)	1.07 ± 0.008	3.45 ± 0.02	−	−	1	8.8	0.04	1
	P_{ln} ($i=5$)	1.07 ± 0.008	3.45 ± 0.02	0.85 ± 0.009	0.94 ± 0.01	0.65 ± 0.03	5.6	0.02	1
3.0 Okinawa	P_{lw} ($i=2$)	1.44 ± 0.02	1.77 ± 0.02	1.40 ± 0.03	2.57 ± 0.04	0.58 ± 0.01	3.8	0.01	1
	P_{pow} ($i=3$)	1.41 ± 0.02	1.63 ± 0.01	2.22 ± 0.04	0.55 ± 0.01	0.79 ± 0.01	8.8	0.07	0.73
	P_{gam} ($i=4$)	1.41 ± 0.02	1.63 ± 0.01	−	−	1	17	0.1	0.34
	P_{ln} ($i=5$)	1.41 ± 0.02	1.63 ± 0.01	0.20 ± 0.003	0.72 ± 0.003	0.04 ± 0.03	5.4	0.04	1
2.5 Okinawa	P_{lw} ($i=2$)	1.72 ± 0.02	1.14 ± 0.008	1.25 ± 0.01	1.68 ± 0.01	0.47 ± 0.006	2.1	0.007	1
	P_{pow} ($i=3$)	1.90 ± 0.05	1.01 ± 0.008	2.84 ± 0.04	0.51 ± 0.005	0.58 ± 0.02	9.2	0.04	1
	P_{gam} ($i=4$)	1.90 ± 0.05	1.01 ± 0.008	1.07 ± 0.02	0.94 ± 0.02	0.99 ± 0.04	22	0.14	0.04
	P_{ln} ($i=5$)	−	−	−0.20 ± 0.004	0.52 ± 0.005	0	12	0.07	0.61
2.0 Okinawa	P_{lw} ($i=2$)	1.78 ± 0.03	0.76 ± 0.009	1.16 ± 0.01	1.44 ± 0.008	0.41 ± 0.008	2.0	0.007	1
	P_{pow} ($i=3$)	2.57 ± 0.10	0.77 ± 0.007	3.61 ± 0.05	0.48 ± 0.003	0.40 ± 0.02	7.0	0.03	1
	P_{gam} ($i=4$)	2.57 ± 0.10	0.77 ± 0.007	1.09 ± 0.03	0.68 ± 0.03	0.96 ± 0.05	25	0.14	0.18
	P_{ln} ($i=5$)	−	−	−0.41 ± 0.005	0.39 ± 0.008	0	15	0.08	0.78

Table 4. Interoccurrence time statistics of earthquakes in Okinawa region. The error bars mean the 95% confidence level of fit.

However, the fitting accuracy of the Weibull distribution becomes worse with a gradual decrease in m_c. We now propose a possible explanation which states that "the interoccurrence time distribution can be described by the superposition of the Weibull distribution and another distribution, hereafter referred to as the distribution $P_X(\tau)$,

$$P(\tau) = p \times \text{Weibull distribution} + (1-p) \times \text{distribution X}$$
$$= p \times P_w(\tau) + (1-p) \times P_X(\tau) \qquad (14)$$

where p is a parameter in the range, $0 \leq p \leq 1$ and stands for the ratio of P_w divided by $P(\tau)$. The interoccurrence time distribution obeys the Weibull distribution for $p = 1$. On the other hand, it follows the distribution $P_X(\tau)$ for $p = 0$. Here, the log-Weibull distribution, the power law distribution, the gamma distribution, and the log normal distribution are candidates for the distribution $P_X(\tau)$.

Next we shall explain the parameter estimation procedures;

(A); the optimal parameters are estimated so as to minimize the differences between the data and the test function by varying five parameters, $\alpha_1, \beta_1, \alpha_i, \beta_i$ and p.

m_c Region	distribution X	Weibull distribution α_1	β_1 [day]	Distribution X α_i	β_i [day]	p	RMS test rms [$\times 10^{-3}$]	KS test DKS	Q
4.5 Chuetsu	$P_{lw}\ (i=2)$	0.75 ± 0.01	27.6 ± 0.40	$-$	$-$	1	21	0.06	0.94
	$P_{pow}\ (i=3)$	0.75 ± 0.01	27.6 ± 0.40	$-$	$-$	1	21	0.06	0.94
	$P_{gam}\ (i=4)$	0.75 ± 0.01	27.6 ± 0.40	$-$	$-$	1	21	0.06	0.94
	$P_{ln}\ (i=5)$	0.75 ± 0.01	27.6 ± 0.40	$-$	$-$	1	21	0.06	0.94
4.0 Chuetsu	$P_{lw}\ (i=2)$	0.81 ± 0.01	10.6 ± 0.13	$-$	$-$	1	17	0.03	1
	$P_{pow}\ (i=3)$	0.81 ± 0.01	10.6 ± 0.13	$-$	$-$	1	17	0.03	1
	$P_{gam}\ (i=4)$	0.81 ± 0.01	10.6 ± 0.13	$-$	$-$	1	17	0.03	1
	$P_{ln}\ (i=5)$	0.81 ± 0.01	10.6 ± 0.13	$-$	$-$	1	17	0.03	1
3.5 Chuetsu	$P_{lw}\ (i=2)$	0.89 ± 0.006	4.84 ± 0.02	2.08 ± 0.06	8.70 ± 0.19	0.93 ± 0.02	6.1	0.03	1
	$P_{pow}\ (i=3)$	0.89 ± 0.006	4.84 ± 0.02	1.66 ± 0.04	0.62 ± 0.05	0.98 ± 0.009	6.3	0.03	1
	$P_{gam}\ (i=4)$	0.89 ± 0.006	4.84 ± 0.02	$-$	$-$	1	6.9	0.04	1
	$P_{ln}\ (i=5)$	0.89 ± 0.006	4.84 ± 0.02	1.12 ± 0.02	1.11 ± 0.02	0.90 ± 0.04	6.2	0.03	1
3.0 Chuetsu	$P_{lw}\ (i=2)$	1.06 ± 0.02	2.00 ± 0.06	1.90 ± 0.16	5.13 ± 0.39	0.82 ± 0.03	3.9	0.012	1
	$P_{pow}\ (i=3)$	1.09 ± 0.008	2.14 ± 0.01	1.98 ± 0.04	0.52 ± 0.02	0.92 ± 0.009	5.0	0.02	1
	$P_{gam}\ (i=4)$	1.09 ± 0.008	2.14 ± 0.01	$-$	$-$	1	6.5	0.03	1
	$P_{ln}\ (i=5)$	1.09 ± 0.008	2.14 ± 0.01	0.39 ± 0.008	0.91 ± 0.02	0.63 ± 0.03	3.7	0.014	1
2.5 Chuetsu	$P_{lw}\ (i=2)$	1.48 ± 0.02	1.21 ± 0.009	1.19 ± 0.02	1.91 ± 0.02	0.62 ± 0.008	2.5	0.01	1
	$P_{pow}\ (i=3)$	1.56 ± 0.03	1.16 ± 0.009	2.53 ± 0.04	0.49 ± 0.008	0.71 ± 0.01	6.4	0.03	1
	$P_{gam}\ (i=4)$	1.56 ± 0.03	1.16 ± 0.009	1.03 ± 0.009	1.09 ± 0.03	0.99 ± 0.04	15	0.01	0.38
	$P_{ln}\ (i=5)$	$-$	$-$	-0.11 ± 0.003	0.64 ± 0.004	0	5.1	0.04	1
2.0 Chuetsu	$P_{lw}\ (i=2)$	1.85 ± 0.03	0.80 ± 0.009	1.18 ± 0.02	1.43 ± 0.01	0.50 ± 0.01	2.2	0.007	1
	$P_{pow}\ (i=3)$	2.46 ± 0.09	0.78 ± 0.007	3.51 ± 0.06	0.48 ± 0.004	0.47 ± 0.03	8.2	0.02	1
	$P_{gam}\ (i=4)$	2.46 ± 0.09	0.78 ± 0.007	1.19 ± 0.05	0.69 ± 0.02	0.97 ± 0.05	2.6	0.12	0.46
	$P_{ln}\ (i=5)$	$-$	$-$	-0.40 ± 0.005	0.40 ± 0.008	0	14	0.06	0.99

Table 5. Interoccurrence time statistics of earthquakes in Chuetsu area. The error bars mean the 95% confidence level of fit.

If there is a parameter, where C_v, the ratio of the standard deviation divided by the mean for a parameter exceeds 0.1, another estimation procedure, (B), is performed.

(B); the Weibull parameters, α_1 and β_1, and the parameters of $P_X(\tau)$, α_i and β_i, are optimized dependently and then p is estimated.

According to those procedures (A) and (B), we obtain the fitting of results of $P(\tau)$. The results for Okinawa and Chuetsu region are listed in Table 4 and 5. We assume that the Weibull distribution is a fundamental distribution, because p becomes unity for large m_c, which means that the effect of the distribution $P_X(\tau)$ is negligible. As observed in Table 4 and 5, the log-Weibull distribution is the most suitable distribution for the distribution $P_X(\tau)$ according to the two goodness-of-fit tests. Thus, we find that the interoccurrence times distribution can be described by the superposition of the Weibull distribution and the log-Weibull distribution, namely,

$$P(\tau) = p \times \text{Weibull distribution} + (1-p) \times \text{log-Weibull distribution},$$
$$= p \times P_w + (1-p) \times P_{lw}, \tag{15}$$

$P(\tau)$ is controlled by five parameters, $\alpha_1, \alpha_2, \beta_1, \beta_2$, and p.

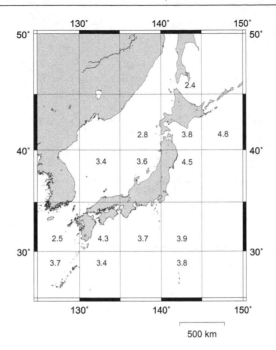

Fig. 7. The crossover magnitude from the superposition regime to the (pure) Weibull regime, denoted m_c^{**}, map around Japan.

It has been shown that the interoccurrence time distribution of earthquakes with large mc obeys the Weibulldistribution with the exponent $\alpha_1 < 1$. As shown in Tables 4 and 5, we stress the point that the distribution function of the interoccurrence time changes by varying m_c. This indicates that the interoccurrence time statistics basically contains both Weibull and log-Weibull statistics, and a dominant distribution function is changed according to the ratio p. In this case, the dominant distribution of the interoccurrence time changes from the log-Weibull distribution to the Weibull distribution when m_c is increased. Thus, the interoccurrence time statistics exhibit transition from the Weibull regime to the log-Weibull regime. The crossover magnitude from the superposition regime to the Weibull regime, denoted by m_c^{**}, depends on the spatial area. We demonstrate the values of m_c^{**} for each regime in Figure 7. As can be seen from the figure, m_c^{**} depends on the region and ranges from 2.4 (140°–145°E and 45°–50°N) to 4.8 (145°–150°E and 40°–45°N). Comparing Figure 1 (a) and Figure 7, we have found that the Weibull - log-Weibull transition occurs in all region, where we analyzed.

3.2 Interoccurrence time statistics in Southern California

Second, we analyze the interoccurrence time statistics using the SCEDC data. The cumulative distributions of interoccurrence time for $m_c = 4.0$ and $m_c = 2.0$ are shown in Figure 8 (a) and (b), respectively. By the rms test and KS test, we confirmed that the Weibull distribution is preferred for large m_c ($m_c = 4.0$) [see Table 6], which is the same result as that from JMA

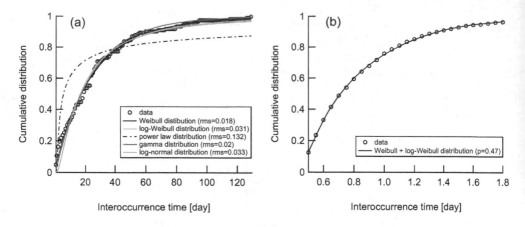

Fig. 8. Cumulative distribution of interoccurrence time in Southern California region different m_c and distribution functions. (a) $m_c = 4.0$ and (b) $m_c = 2.0$.

m_c Region	distribution X	Distribution X α_i	β_i [day]	RMS test rms [$\times 10^{-3}$]	KS test DKS	Q
	P_l $(i = 1)$	0.91 ± 0.01	25.4 ± 0.21	18	0.07	0.56
	P_{lw} $(i = 2)$	3.66 ± 0.07	48.2 ± 0.68	31	0.14	0.11
4.0	P_{pow} $(i = 3)$	1.44 ± 0.02	1.15 ± 0.15	133	0.49	1.6×10^{-9}
SCEDC	P_{gam} $(i = 4)$	0.97 ± 0.004	25.6 ± 0.22	20	0.08	0.36
	P_{ln} $(i = 5)$	2.79 ± 0.02	1.10 ± 0.02	33	0.13	0.02

Table 6. Results of rms value, DKS, and Q for different distribution functions in the case of $m_c = 4.0$ for Southern California earthquakes. The error bars mean the 95% confidence level of fit.

data. Unfortunately, the fitting accuracy of the Weibull distribution gets worse by decreasing a threshold m_c. We propose the following hypothesis, "the interoccurrence time distribution can be described by the superposition of the Weibull distribution and the distribution $P_X(\tau)$" which is the same in 3.1. As shown in Table 7, the log-Weibull distribution is the most suitable for the distribution $P_X(\tau)$, because the smallest rms-value, the smallest DKS value and the largest Q value can be obtained. Therefore, we find that the Weibull - log-Weibull transition shows in Southern California earthquakes. The crossover magnitude m_c^{**} is estimated to be 3.3.

3.3 Interoccurrence time statistics in Taiwan

Finally, the TCWB data was analyzed to investigate the interoccurrence time statistics in Taiwan. Figure 9 shows the cumulative distribution of interoccurrence time for $m_c = 4.5$ and $m_c = 3.0$, respectively. For large m_c, the Weibull distribution is preferred on the basis of the rms and KS test [see Table 8]. As the threshold of magnitude m_c decreases, the fitting accuracy of the Weibull distribution is getting worse, as is common in JMA and SCEDC. According to a hypothesis that the interoccurrence time distribution can be described by the superposition of

m_c Region	distribution X	Weibull distribution α_1	β_1 [day]	Distribution X α_i	β_i [day]	p	RMS test rms [$\times10^{-3}$]	KS test DKS	Q
	P_{lw} ($i=2$)	0.91 ± 0.01	25.4 ± 0.21	−	−	1	18	0.07	0.56
4.0	P_{pow} ($i=3$)	0.91 ± 0.01	25.4 ± 0.21	−	−	1	18	0.07	0.56
SCEDC	P_{gam} ($i=4$)	0.91 ± 0.01	25.4 ± 0.21	−	−	1	18	0.07	0.56
	P_{ln} ($i=5$)	0.91 ± 0.01	25.4 ± 0.21	−	−	1	18	0.07	0.56
	P_{lw} ($i=2$)	0.83 ± 0.006	9.29 ± 0.05	−	−	1	11	0.03	1
3.5	P_{pow} ($i=3$)	0.83 ± 0.006	9.29 ± 0.05	−	−	1	11	0.03	1
SCEDC	P_{gam} ($i=4$)	0.83 ± 0.006	9.29 ± 0.05	−	−	1	11	0.03	1
	P_{ln} ($i=5$)	0.83 ± 0.006	9.29 ± 0.05	−	−	1	11	0.03	1
	P_{lw} ($i=2$)	1.01 ± 0.01	3.08 ± 0.04	1.37 ± 0.07	3.49 ± 0.19	0.80 ± 0.01	4.1	0.01	1
3.0	P_{pow} ($i=3$)	0.98 ± 0.008	2.85 ± 0.02	1.85 ± 0.03	0.58 ± 0.03	0.91 ± 0.009	6.1	0.03	1
SCEDC	P_{gam} ($i=4$)	−	−	0.99 ± 0.002	2.84 ± 0.01	1	8.2	0.05	1
	P_{ln} ($i=5$)	0.98 ± 0.008	2.85 ± 0.02	0.63 ± 0.009	1.00 ± 0.01	0.60 ± 0.03	5.0	0.03	1
	P_{lw} ($i=2$)	1.32 ± 0.01	1.72 ± 0.01	1.35 ± 0.01	2.40 ± 0.02	0.57 ± 0.005	2.0	0.01	1
2.5	P_{pow} ($i=3$)	1.33 ± 0.02	1.55 ± 0.009	2.27 ± 0.03	0.54 ± 0.008	0.76 ± 0.01	6.4	0.04	0.96
SCEDC	P_{gam} ($i=4$)	1.33 ± 0.02	1.17 ± 0.009	−	−	1	12	0.10	0.04
	P_{ln} ($i=5$)	−	−	0.13 ± 0.002	0.75 ± 0.003	0	4.4	0.04	0.92
	P_{lw} ($i=2$)	1.88 ± 0.03	0.92 ± 0.008	1.15 ± 0.02	1.56 ± 0.01	0.47 ± 0.008	2.7	0.007	1.1
2.0	P_{pow} ($i=3$)	2.18 ± 0.07	0.88 ± 0.008	3.14 ± 0.05	0.49 ± 0.005	0.52 ± 0.02	8.3	0.11	0.88
SCEDC	P_{gam} ($i=4$)	2.18 ± 0.07	0.88 ± 0.008	1.15 ± 0.04	0.80 ± 0.02	0.98 ± 0.05	26	0.13	0.72
	P_{ln} ($i=5$)	−	−	−0.31 ± 0.005	0.46 ± 0.007	0	7.5	0.007	1

Table 7. Interoccurrence time statistics of earthquakes in Southern California area. The error bars mean the 95% confidence level of fit.

the Weibull distribution and another distribution, we investigate the distribution $P_X(\tau)$. As shown in Table 9, the log-Weibull distribution is preferred as the distribution on the basis of the two goodness-of-fit tests. We estimated the crossover magnitude m_c^{**} to be 4.9.

3.4 Brief summary of the interoccurrence time statistics for earthquakes

Taken all together, we clarified that distribution of interoccurrence time is well fitted by the superposition of the Weibull distribution and log-Weibull distribution. For large m_c, $P(\tau)$ obeys the Weibull distribution with $\alpha_1 < 1$, indicating that the occurrence of earthquakes is not a Possion process. When the threshold of magnitude m_c decreases, the ratio of the Weibull distribution of $P(\tau)$ gradually increases. We suggest that the Weibull statistics and log-Weibull statistics coexist in interoccurrence time statistics, where the change of the distribution means the change of a dominant distribution. In this case, the dominant distribution changes from the log-Weibull distribution to the Weibull distribution by increasing the m_c. It follows that the Weibull - log-Weibull transition exists in Japan, Southern California, and Taiwan.

4. Discussion

4.1 Size dependency

To investigate the region-size, L dependency of the Weibull - log-Weibull transition, we change the window size L is varied from $3°$ to $25°$ (17). In (17), we use JMA data we used is from

m_c Region	distribution X	Distribution X α_i	β_i [day]	RMS test rms [$\times 10^{-3}$]	KS test DKS	Q
5.0 TCWB	P_l $(i=1)$	0.86 ± 0.006	15.1 ± 0.08	8.4	0.02	1
	P_{lw} $(i=2)$	3.01 ± 0.06	27.9 ± 0.42	24	0.10	0.24
	P_{pow} $(i=3)$	1.51 ± 0.03	0.93 ± 0.11	111	0.25	3.6×10^{-6}
	P_{gam} $(i=4)$	0.97 ± 0.003	15.3 ± 0.14	16	0.05	0.95
	P_{ln} $(i=5)$	2.23 ± 0.02	1.15 ± 0.02	24	0.08	0.53

Table 8. Results of rms value, DKS, and Q for different distribution functions in the case of $m_c = 5.0$ for Taiwan earthquakes. The error bars mean the 95% confidence level of fit.

m_c Region	distribution X	Weibull distribution α_1	β_1 [day]	Distribution X α_i	β_i [day]	p	RMS test rms [$\times 10^{-3}$]	KS test DKS	Q
5.0 TCWB	P_{lw} $(i=2)$	0.86 ± 0.006	15.0 ± 0.08	–	–	1	8.4	0.02	1
	P_{pow} $(i=3)$	0.86 ± 0.006	15.0 ± 0.08	–	–	1	8.4	0.02	1
	P_{gam} $(i=4)$	0.86 ± 0.006	15.0 ± 0.08	–	–	1	8.4	0.02	1
	P_{ln} $(i=5)$	0.86 ± 0.006	15.0 ± 0.08	–	–	1	8.4	0.02	1
4.5 TCWB	P_{lw} $(i=2)$	0.88 ± 0.004	5.34 ± 0.02	–	–	1	4.3	0.01	1
	P_{pow} $(i=3)$	0.88 ± 0.004	5.34 ± 0.02	–	–	1	4.3	0.01	1
	P_{gam} $(i=4)$	0.88 ± 0.004	5.34 ± 0.02	–	–	1	4.3	0.01	1
	P_{ln} $(i=5)$	0.88 ± 0.004	5.34 ± 0.02	–	–	1	4.3	0.01	1
4.0 TCWB	P_{lw} $(i=2)$	1.00 ± 0.01	2.30 ± 0.04	1.82 ± 0.05	4.06 ± 0.13	0.68 ± 0.01	2.8	0.009	1
	P_{pow} $(i=3)$	1.08 ± 0.01	2.30 ± 0.02	1.95 ± 0.04	0.55 ± 0.02	0.89 ± 0.01	6.9	0.03	1
	P_{gam} $(i=4)$	1.08 ± 0.01	2.30 ± 0.02	–	–	1	10	0.05	0.99
	P_{ln} $(i=5)$	1.08 ± 0.01	2.30 ± 0.02	0.46 ± 0.005	0.92 ± 0.006	0.39 ± 0.02	2.8	0.007	1
3.5 TCWB	P_{lw} $(i=2)$	1.44 ± 0.03	1.29 ± 0.02	1.32 ± 0.05	2.15 ± 0.05	0.61 ± 0.02	4.5	0.01	1
	P_{pow} $(i=3)$	1.52 ± 0.03	1.25 ± 0.01	2.41 ± 0.05	0.50 ± 0.01	0.74 ± 0.02	8.6	0.07	0.94
	P_{gam} $(i=4)$	1.52 ± 0.03	1.25 ± 0.01	–	–	1	18	0.09	0.73
	P_{ln} $(i=5)$	–	–	-0.04 ± 0.003	0.66 ± 0.004	0	5.4	0.03	1
3.0 TCWB	P_{lw} $(i=2)$	2.07 ± 0.06	0.75 ± 0.009	1.18 ± 0.03	1.47 ± 0.01	0.51 ± 0.01	3.2	0.01	1
	P_{pow} $(i=3)$	2.62 ± 0.09	0.78 ± 0.006	3.58 ± 0.08	0.48 ± 0.005	0.53 ± 0.02	7.5	0.02	1
	P_{gam} $(i=4)$	2.62 ± 0.09	0.78 ± 0.006	1.26 ± 0.08	0.66 ± 0.03	0.97 ± 0.05	23	0.10	0.76
	P_{ln} $(i=5)$	–	–	-0.41 ± 0.004	0.38 ± 0.006	0	12	0.05	1
2.5 TCWB	P_{lw} $(i=2)$	5.35 ± 0.41	0.60 ± 0.005	0.86 ± 0.05	1.16 ± 0.007	0.43 ± 0.04	16	0.037	1
	P_{pow} $(i=3)$	5.35 ± 0.41	0.60 ± 0.005	6.54 ± 0.09	0.49 ± 0.001	0.10 ± 0.04	10	0.041	1
	P_{gam} $(i=4)$	5.35 ± 0.41	0.60 ± 0.005	1.69 ± 2.04	0.33 ± 0.61	0.94 ± 0.07	46	0.17	0.41
	P_{ln} $(i=5)$	–	–	-0.58 ± 0.006	0.19 ± 0.01	0	35	0.11	0.87

Table 9. Interoccurrence time statistics of earthquakes in Taiwan area. The error bars mean the 95% confidence level of fit.

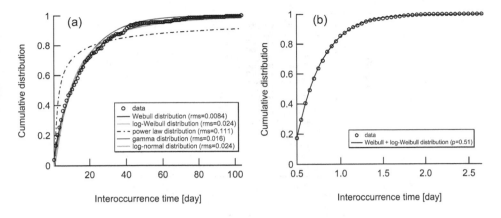

Fig. 9. Cumulative distribution of interoccurrence time for Taiwan for different m_c and distribution function. (a) and (b) represent interoccurrence time when $m_c = 4.5$ and $m_c = 3.0$, respectively.

L	Region	m_c	α_1	β_1 [day]	rms	m_c^{**}
$L = 3$	140°-143° E and 35°-38° N	3.9	0.88 ± 0.01	19.4 ± 0.18	0.011	4.6
$L = 5$	140°-145° E and 35°-40° N	4.0	0.75 ± 0.02	10 ± 0.19	0.014	4.7
$L = 10$	140°-150° E and 35°-45° N	4.2	0.94 ± 0.005	8.36 ± 0.04	0.0077	4.9
$L = 25$	125°-150° E and 25°-50° N	5.0	0.93 ± 0.01	17.8 ± 0.10	0.041	5.7

Table 10. Interoccurrence time statistics for different system size L by analyzing the JMA data from 1st January 2001 to 31st October 2007 (17).

1st January 2001 to 31st October 2007. We use the data covering the region 140°-143° E and 35°-38° N for $L = 3$, 140°-145° E and 35°-40° N for $L = 5$, 140°-150° E and 35°-45° N for $L = 10$, and 125°-150° E and 25°-50° N for $L = 25$. As for $L = 25$, the data covers the whole region of the JMA catalogue. The result of fitting parameters of $P(\tau)$, the crossover magnitude m_c^{**}, and the rms value are listed in Table 10. It is demonstrated that in all the cases, the Weibull exponent α_1 is less than unity and the Weibull - log-Weibull transition appears. m_c^{**} depends on L, namely $m_c^{**} = 3.9$ for $L = 3$, $m_c^{**} = 4.0$ for $L = 5$, $m_c^{**} = 4.2$ for $L = 10$, and $m_c^{**} = 5.0$ for $L = 25$. Therefore we can conclude that the interoccurrence time statistics, namely the Weibull - log-Weibull transition, presented here hold from $L = 3$ to $L = 25$.

4.2 Relation between the m_c^{**} and m_{max}

To study the feature of the Weibull - log-Weibull transition, we summarize our results obtained from 16 different regions (14 regions in Japan, Southern California, and Taiwan.) Interestingly, m_c^{**} is proportional to the maximum magnitude of an earthquake in a region, where we analyzed, denoted here m_{max} [see Figure 10]. We then obtain a region-independent relation between m_c^{**} and m_{max},

$$m_c^{**} / m_{max} = 0.56 \pm 0.08 \qquad (16)$$

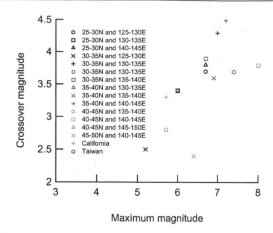

Fig. 10. The crossover magnitude m_c^{**} v.s the maximum magnitude m_{max} in 16 regions (14 region in Japan, Southern California, and Taiwan).

Region	relative plate motion	velocity [mm/yr]	m_c^{**}
Taiwan	PH-EU	71	4.90
East Japan	PA-PH	49	4.07 [1]
West Japan	PH-EU	47	3.74 [2]
California	PA-NA	47	3.30

Table 11. List of the crossover magnitude, m_c^{**} and the plate velocity (32; 33). The notation of PH, EU, PA, and NA represent PHilippine Sea plate, EUrasian plate, PAcific plate, and North American plate, respectively.

1 We take an average using three regions; 25°–30°N, 140°–145°E ($m_c^{**} = 3.8$), 30°–35°N, 140°–145°E ($m_c^{**} = 3.9$), 35°–40°N, 140°–145°E ($m_c^{**} = 4.5$).

2 We take an average using five regions; 25°–30°N, 125°–130°E ($m_c^{**} = 3.7$), 25°–30°N, 130°–135°E ($m_c^{**} = 3.4$), 30°–35°N, 130°–135°E ($m_c^{**} = 4.3$), 30°–35°N, 135°–140°E ($m_c^{**} = 3.7$), 35°–40°N, 135°–140°E ($m_c^{**} = 3.6$).

This relation can be useful to interpret the Weibull - log-Weibull transition of geophysical meaning.

4.3 Interpretation of the Weibull - log-Weibull transition

Although the scaled crossover magnitudes m_c**/m_{max} is region-independent, the crossover magnitude m_c^{**} from the superposition regime to the pure Weibull regime probably depends on the tectonic region (Figure 7). To investigate the Weibull - log-Weibull transition further, we consider the plate velocity with m_c^{**}, which can shed light on the geophysical implication of the region-dependent m_c^{**}. As shown in Table 11, m_c^{**} is on the average proportional to the plate velocity. That means that the maximum magnitude m_{max} for a tectonic region is more or less proportional to the plate velocity since $m_c^{**}/0.56 = m_{max}$. Such an interesting consequence is reminiscent of the early study by Ruff and Kanamori (1980) (34). They showed a relation that the magnitude of characteristic earthquake occurred in the subduction-zone, M_w is directly proportional to the plate-velocity, V, and is directly inversely proportional to plate-age T, namely,

$$M_w = -0.000953T + 1.43V + 8.01. \tag{17}$$

The relation $m_c**/0.56 = m_{max}$ can thus be explained on the basis of their early observation about the velocity-dependence of the characteristic earthquake magnitude. The physical interpretation of the Weibull - log-Weibull transition remains open. However, it might suggest that the occurrence mechanism of earthquake could probably depend on its magnitude then, inevitably, the distribution of the interoccurrence time statistics changes as the threshold of magnitude m_c is varied. It is well known that the Weibull distribution for life-time of materials can be derived in the framework of damage mechanics (4; 24; 35–37). Our present results thus suggest that larger earthquakes might be caused by the damage mechanism driven by the plate motion, whereas the effect of the plate-driven damaging process might become minor for smaller earthquakes. Hence, the transition from the Weibull regime to the log-Weibull regime could be interpreted from the geophysical sense as the decrement of the plate-driven damaging mechanics.

4.4 A universal relation and intrinsic meanings of the Gutenberg-Richter parameter

Here we consider the interrelation between the Gutenberg-Richter law, denoted in this subsection $P(m) \propto e^{-bm}$ and the Weibull distribution for the interoccurrence time $(P(\tau) \propto t^{-\alpha-1} \cdot e^{(\tau/\beta)^{\alpha}})$. We assume that these two statistics are correct over wide ranges, and the parameters (α, β) are depending on the magnitude, i.e., $\alpha(m)$ and $\beta(m)$, then the following relation is easily obtained from the calculation of the mean interoccurrence time between two earthquakes whose magnitude is larger than m ,

$$\beta(m_1)e^{-bm_1}\Gamma\left(1+\frac{1}{\alpha_1}\right) = \beta(m_2)e^{-bm_2}\Gamma\left(1+\frac{1}{\alpha_2}\right), \tag{18}$$

where m_1 and m_2 are arbitrary values of m. This implies that the quantity defined by $\beta(m_1)e^{-bm_1}\Gamma\left(1+\frac{1}{\alpha_1}\right)$ is a universal constant when we consider the local earthquakes in a relatively small area.

One of the most important results derived from equation (18) is that the GR parameter b is determined by two parameters, in other words, the parameters (α, β) depend on the magnitude m as well as on the GR parameter b,

$$\alpha = f_\alpha(m, b)$$
$$\beta = f_\beta(m, b), \tag{19}$$

where the functional forms of f_α and f_β characterize the time series of earthquakes under consideration.

It is difficult to determine those forms completely from any seismological relations known so far, but it is possible for us to obtain the universal aspects of f_α and f_β by a perturbational

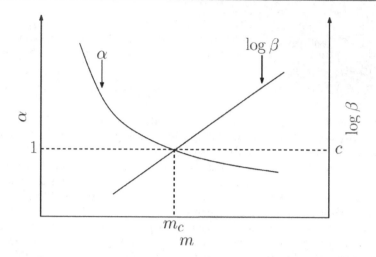

Fig. 11. Schematic picture of universal behavior of $f_\alpha(m, b)$ and $f_\beta(m, b)$ near $m \simeq m_c$.

approach. Here we consider a particular solution of equation (19) which satisfies the following conditions; $f_\beta(m, b) = \exp\left[b(m - m_c) + c\right]$ and $f_\alpha(m_c, b)$, namely, the characteristic time β is a exponentilally increasing function of m, and the interoccurrence time distribution is an exponential one ($\alpha = 1$) at $m = m_c$, where b' and c are constant parameters. By use of this simplification, equation (18) is rewritten by putting $m_1 = m$ and $m_2 = m$,

$$(b' - b)(m - m_c) = -\log \Gamma\left(1 + \frac{1}{\alpha(m)}\right)$$

$$\cong \frac{1}{2}\Delta - \frac{3}{4}(\Delta)^2 + \cdots, \quad (\Delta = \alpha(m - 1). \tag{20}$$

Here we used the Taylor expansion near $m \cong m_c$ (i.e., $\alpha(m) \cong \alpha(m_c)$). Figure 11 shows the schematic result of equation (20). One can see that the universal relation is recognized in many cases treated in this chapter (section 4.3.3.), though the exponential growth of β, $\log \beta(m) \simeq b'(m - m_c) + c$ is a little bit accelerated.

We have to remind that the solution mentioned above is not unique, but many other solutions for equation (19) are possible under the universal relation of equation (18). Further details will be studied in our forthcoming paper (38).

4.5 Comparison with previous works

Finally, we compared our results with those of previous studies. The unified scaling law shows a generalized gamma distribution [see in equations (2), and (3)] which is approximately the gamma distribution, because δ in Corral's paper (6) is close to unity ($\delta = 0.98 \pm 0.05$). For a long time domain, this distribution decays exponentially, supporting the view that an earthquake is a Poisson process. However, we have demonstrated that the Weibull distribution is more appropriate than the gamma distribution on the basis of two goodness-fit-tests. In addition, for large m_c, the distribution in a long time domain is similar

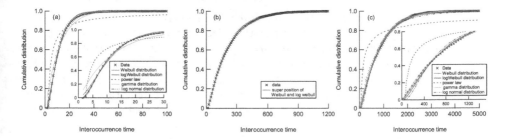

Fig. 12. Interoccurrence time statistics for different magnitude m_c by analyzing catalogue produced by the two-dimensional spring-block model (29).

to the stretched exponential distribution because α_1 is less than unity, suggesting that an occurrence of earthquake is not a Poisson process but has a memory. We provide the first evidence that the distribution changes from the Weibull to log-Weibull distribution by varying m_c, i.e., the Weibull - log-Weibull transition. Recently, Abaimov *et al.* showed that the recurrence time distribution is also well-fitted by the Weibull distribution (4) rather than the Brownian passage time (BPT) distribution (23) and the log normal distribution. Taken together, we infer that both the recurrence time statistics and the interoccurrence time statistics show the Weibull distribution.

In this chapter, we propose a new insight into the interoccurrence time statistics, stating that the interoccurrence statistics exhibit the Weibull - log-Weibull transition by analyzing the different tectonic settings, JMA, SCEDC, and TCWB. This stresses that the distribution function can be described by the superposition of the Weibull distribution and the log-Weibull distribution, and that the predominant distribution function changes from the log-Weibull distribution to the Weibull distribution as m_c is increased. Note that there is a possibility that a more suitable distribution might be found instead of the log-Weibull distribution. Furthermore, the Weibull - log-Weibull transition can be extracted more clearly by analyzing synthetic catalogs produced by the spring-block model [see Figure 12] (29).

5. Conclusion

In conclusion, we have proposed a new feature of interoccurrence time statistics by analyzing the Japan (JMA), Southern California, (SCEDC), and Taiwan (TCWB) for different tectonic conditions. We found that the distribution of the interoccurrence time can be described clearly by the superposition of the Weibull distribution and the log-Weibull distribution. Especially for large earthquakes, the interoccurrence time distribution obeys the Weibull distribution with the exponent $\alpha_1 < 1$, indicating that a large earthquake is not a Poisson process but a phenomenon exhibiting a long-tail distribution. As the threshold of magnitude m_c increases, the ratio of the Weibull distribution in the interoccurrence time distribution p gradually increases. Our findings support the view that the Weibull statistics and log-Weibull statistics coexist in the interoccurrence time statistics. We interpret the change of distribution function as the change of the predominant distribution function; the predominant distribution changes from the log-Weibull distribution to the Weibull distribution when m_c is increased. Therefore, we concluded that the interoccurrence time statistics exhibit a Weibull - log-Weibull transition. We also find the region-independent relation, namely, $m_c^{**}/m_{max} = 0.56 \pm 0.08$. In addition, the crossover magnitude m_c^{**} is proportional to the plate velocity, which is consistent with an earlier observation about the velocity-dependence of the characteristic earthquake magnitude (34). Although the origins of both the log-Weibull distribution and the Weibull - log-Weibull transition remain open questions, we suggest the change in the distribution from the log-Weibull distribution to the Weibull distribution can be considered as the enhancement in the plate-driven damaging mechanics. We believe that this work is a first step toward a theoretical and geophysical understanding of this transition.

6. Acknowledgments

The authors thank the editor of this book and the publisher for giving us the opportunity to take part in this book project. We would like to thank the JMA, SCEDC and TCWB for allowing us to use the earthquake data. The effort of the Taiwan Central Weather Bureau for maintaining the CWB Seismic Network is highly appreciated. TH is supported by the Japan Society for the Promotion of Science (JSPS) and the Earthquake Research Institute cooperative research program at the University of Tokyo. This work is partly supported by the Sasagawa Scientific Research Grant from The Japan Science Society by the Waseda University.. CCC is also grateful for research supports from the National Science Council (ROC) and the Department of Earth Sciences at National Central University (ROC).

7. References

[1] I. G. Main, "Statistical physics, seismogenesis, and seismic hazard", *Review of Geophysics*, vol. 34, pp. 433-462, 1996.

[2] B. Gutenberg and C. F. Richter, "Magnitude and energy of earthquakes", *Annals of Geophysics*, vol. 9, pp.1-15, 1956.

[3] F. Omori, "On the after-shocks of earthquakes", *Journal of the College of Science, Imperial University of Tokyo*, vol. 7, pp. 111-200, 1894.

[4] S. G. Abaimov, D. L. Turcotte, R. Shcherbakov, and J. B. Rundle, "Recurrence and interoccurrence behavior of self-organized complex phenomena", *Nonlinear Processes in Geophysics*, vol. 14, pp. 455-464, 2007.

[5] P. Bak, K. Christensen, L. Danon, and T. Scanlon, "Unified Scaling Law for Earthquakes", *Physical Review Letters*, vol. 88, 178501, 2002.

[6] A. Corral, "Long-Term Clustering, Scaling, and Universality in the Temporal Occurrence of Earthquakes", *Physical Review Letters*, vol. 92, 108501, 2004.

[7] R. Shcherbakov, G. Yakovlev, D. L. Turcotte, and J. B. Rundle, "Model for the Distribution of Aftershock Interoccurrence Times", *Physical Review Letters*, vol. 95, 218501, 2005.

[8] A. Saichev and D. Sornette, ""Universal" Distribution of Inter-Earthquake Times Explained", *Physical Review Letters*, vol. 97, 078501, 2006.

[9] A. Saichev and D. Sornette, "Theory of earthquake recurrence times", *Journal of Geophysics Research*, vol. 112, B04313, 2007.

[10] S. Abe and N. Suzuki, "Scale-free statistics of time interval between successive earthquakes", *Physica A*, vol. 350, pp. 588-596, 2005.

[11] J. K. Gardner, L. Knopoff, Is the sequence of earthquakes in southern California with aftershocks removed, Poissonian?, *Bulletin of the Seismological Society of America*, vol. 64, pp.1363-1367, 1974.

[12] B. Enescu, Z. Struzik, and K. Kiyono, On the recurrence time of earthquakes: insight from Vrancea (Romania) intermediate-depth events, *Geophysical Journal International*, vol. 172, pp. 395-404, 2008.

[13] A. Bunde, J. F. Eichner, J. W. Kantelhardt, and S. Havlin, Long-Term Memory: A Natural Mechanism for the Clustering of Extreme Events and Anomalous Residual Times in Climate Records, *Physical Review Letters*, vol. 94, 048701, 2005.

[14] V. N. Livina, S. Havlin, and A. Bunde, Memory in the Occurrence of Earthquakes, *Physical Review Letters*, vol. 95, 208501, 2005.

[15] S. Lennartz, V. N. Livina, A. Bunde, and S. Havlin, Long-term memory in earthquakes and the distribution of interoccurrence times, *Europhysics Lettters*, vol. 81, 69001, 2008.

[16] T. Akimoto, T. Hasumi, and Y. Aizawa, "Characterization of intermittency in renewal processes: Application to earthquakes", *Physical Review E* vol. 81, 031133, 2010.

[17] T. Hasumi, T. Akimoto, and Y. Aizawa, "The Weibull - log-Weibull distribution for interoccurrence times of earthquakes", *Physica A*, vol. 388, pp. 491-498, 2009.

[18] T. Hasumi, C. Chen, T. Akimoto, and Y. Aizawa, "The Weibull - log-Weibull transition of interoccurrence times for synthetic and natural earthquakes", *Techtonophysics*, vol. 485, pp. 9-16, 2010.

[19] Japan Meteorological Agency Earthquake Catalog
: http://wwweic.eri.u-tokyo.ac.jp/db/jma1.

[20] Southern California Earthquake Data Center: http://www.data.scec.org

[21] Taiwan Central Weather Bureau: http://www.cwb.gov.tw/

[22] T. Huillet and H. F. Raynaud, "Rare events in a log-Weibull scenario-Application to earthquake magnitude data", *The European Physical Journal B*, vol. 12, pp. 457-469, 1999.

[23] M. V. Matthews, W. L. Ellsworth, and P. A. Reasenberg, "A brownian model for recurrent earthquakes", *Bulletin of the Seismological Society of America*, vol. 92, pp. 2233-2250, 2002.

[24] W. Weibull, A statistical distribution function of wide applicability, *Journal of Applied Mathematics*, vol. 18, pp. 293-297, 1951.

[25] Y. Hagiwara, "Probability of earthquake occurrence as obtained from a Weibull distribution analysis of crustal strain", *Tectonophys*, vol. 23, pp. 313-318, 1974.

[26] W. H. Bakun, B. Aagard, B. Dost, *et al.*, "Implications for prediction and hazard assessment from the 2004 Parkfield earthquake", *Nature*, vol. 437, pp. 969-974, 2005.

[27] S. G. Abaimov, D. L. Turcotte, and J. B. Rundle, "Recurrence-time and frequency-slip statistics of slip events on the creeping section of the San Andreas fault in central California", *Geophysical Journal International*, vol.170, pp. 1289-1299, 2007.

[28] K. Z. Nanjo, D. L. Turcotte, and R. Shcherbakov, "A model of damage mechanics for the deformation of the continental crust", *Journal of Geophysics Research*, vol. 110, B07403, 2005.

[29] T. Hasumi, T. Akimoto, and Y. Aizawa, "The Weibull - log-Weibull transition of the interoccurrence statistics in the two-dimensional Burridge-Knopoff earthquake model", *Physica A*, vol. 388, pp. 483-490, 2009.

[30] G. Yakovlev, D. L. Turcotte, J. B. Rundle, and P. B. Rundle, "Simulation-Based Distributions of Earthquake Recurrence Times on the San Andreas Fault System", *Bulletin of the Seismological Society of America*, vol. 96, pp. 1995-2007, 2006.

[31] W. H. Press, S. A. Teukolsky, W. T. Vetterling, and B. P. Flannery, *Numerical Recipes in C 2nd edition*, Cambridge University Press, Cambridge, 1995.

[32] T. Seno, S. Seth, and E. G. Alice, "A model for the motion of the Philippine Sea plate consistent with NUVEL-1 and geological data", *Journal of Geophysics Research*, vol 98, pp. 17941-17948, 1993.

[33] C. M. R. Fowler, *The Solid Earth: An Introduction to Global Geophysics*, Cambridge University Press, New York, 1990.

[34] L. Ruff and H. Kanamori, "Seismicity and the subduction process", *Physics of the Earth and Planetary Interiors*, vol. 23, pp. 240-252, 1980.

[35] T. Wong, R. H. C. Wong, K. T. Chau, and, C. A. Tang, "Microcrack statistics. Weibull distribution and micromechanical modeling of compressive failure in rock", *Mechanics of Materials*. vol. 38, pp. 664-681, 2006.

[36] A. Ghosh, "A FORTRAN program for fitting Weibull distribution and generating samples", *Computational Geosciences*, vol. 25, pp. 729-738, 1999.

[37] D. L. Turcotte, W. I. Newman, R. Shcherbakov, "Micro and macroscopic models of rock fracture", *Geophysical Journal International*, vol. 152, pp. 718-728, 2003.

[38] Y. Aizawa and T. Hasumi, in preparation (2011).

2

Change Point Analysis in Earthquake Data

Ayten Yiğiter

Hacettepe University, Faculty of Science, Department of Statistics Beytepe-Ankara
Turkey

1. Introduction

Earthquake forecasts are very important in human life in terms of estimating hazard and managing emergency systems. Defining of earthquake characteristics plays an important role in these forecasts. Of these characteristics, one is the frequency distribution of earthquakes and and the other is the magnitude distribution of the earthquakes. Each statistical distribution has many parameters describing the actual distribution.

There are various statistical distributions used to model the earthquake numbers. As is well known, these are binomial, Poisson, geometric and negative binomial distributions. It is generally assumed that earthquake occurrences are well described by the Poisson distribution because of its certain characteristics (for some details, see Kagan, 2010; Leonard & Papasouliotis, 2001). In their study, Rydelek & Sacks (1989) used the Poisson distribution of earthquakes at any magnitude level. The Poisson distribution is generally used for earthquakes of a large magnitude, and the earthquake occurrences with time/space can be modeled with the Poisson process in which, as is known, the Poisson distribution is one that counts the events that have occurred over a certain period of time.

There is a significant amount of research on the change point as applied to earthquake data. Amorese (2007) used a nonparametric method for the detection of change points in frequency magnitude distributions. Yigiter & İnal (2010) used earthquake data for their method developed for the estimator of the change point in Poisson process. Aktaş et al. (2009) investigated a change point in Turkish earthquake data. Rotondi & Garavaglia (2002) applied the hierarchical Bayesian method for the change point in data, taken from the Italian NT4.1.1 catalogue.

Recently, much research in the literature has focused on whether there is an increase in the frequency of earthquake occurrences. It is further suggested that any increase in the frequency of earthquakes, in some aspects, is due to climate change in the world. There is considerable debate on whether climate change really does increase the frequency of natural disasters such as earthquakes and volcano eruptions. In many studies, it is emphasized that there is serious concern about impact of climate change on the frequencies of hazardous events (Peduzzi, 2005; Lindsey, 2007; Mandewille, 2007 etc.). In Peduzzi's study (2005), there are some indicators about increasing number of the earthquakes especially affecting human settlements, and it is also reported that there is an increase in the percentage of earthquakes affecting human settlements from 1980 onwards. The change point analysis can be used to study the increase or decrease in the frequency of the earthquake occurrences.

The change point analysis is a very useful statistical tool to detect an abrupt or a structural change in the characteristic of the data observed sequentially or chronologically. Many different statistical methods are available to detect the change point in the distribution of a sequence of random variables (Smith, 1975; Hinkley, 1970; Boudjellaba et al., 2001 in many others). Multiple change points have also been investigated for many years (Hendersen & Matthews, 1993; Yao, 1993; Chen & Gupta, 1997, among others.).

In this study, we aimed to find an evidence of increased or decreased in world seismic activities after 1901. Earthquake frequencies are modeled by the Poisson distribution as the most commonly used discrete distribution. We modeled the earthquakes in the world with the Poisson process since the number of earthquakes is counted with Poisson distribution. We investigated any abrupt change point(s) in parameters of the model using the frequentist (maximum likelihood) and Bayesian method. When the magnitude of the earthquakes is taken into account in the process, it is then called the compound Poisson process. We also investigated a change in the magnitudes of the world earthquakes. For this purpose, Poisson process and compound Poisson process are introduces in Section 2. The frequentist and Bayesian methods used for change point estimates are explained in Section 3. Worldwide earthquake data and change point analysis of this data are given Section 4.

2. Poisson process

The Poisson process has an important place in stochastic processes. It is a Markov chain with a continuous parameter.

A stochastic process $\{N_t, t \geq 0\}$ is said to be a homogeneous Poisson process if

i. $\{N_t, t \geq 0\}$ has stationary independent increments

ii.   for any times s and t such that s<t the number of events occurred in time interval (s, t) has Poisson distribution with parameter $\lambda(t-s)$.

In the homogeneous Poisson Process, events occur independently throughout time. The number of events occurred time interval (0, t] has Poisson distribution with parameter λt,

$$P(N_t = i) = e^{-\lambda t} \frac{(\lambda t)^i}{i!}, \qquad i = 0, 1, \ldots \tag{1}$$

$$= 0, \qquad\qquad \text{otherwise}$$

where λ is called the occurrence rate of the event in unit time. The occurrence rate is assumed to be a constant throughout the process.

Let $S_0, S_1, S_2, \ldots$ be occurrence times of the events where $S_0 = 0$ and $T_1 = S_1 - S_0$, $T_2 = S_2 - S_1, \ldots$ be time intervals between the events. $T_1, T_2, \ldots$ are independent identically exponential random variables with parameter λ. The distribution of the random variables $T_1, T_2, \ldots$ is:

$$f(t) = \lambda e^{-\lambda t}, \quad t \geq 0$$
$$= 0, \qquad t < 0 \tag{2}$$

2.1 Compound poisson process

A stochastic process $\{X_t, t \geq 0\}$ is said to be a compound Poisson process if it can be presented, for $t \geq 0$, by

$$X_t = \sum_{i=1}^{N_t} Y_i \tag{3}$$

in which $\{N_t, t \geq 0\}$ is a Poisson process, and $\{Y_i, i = 1, 2, \ldots\}$ are independent identically distributed (i.i.d.) random variables. The compound Poisson distribution has stationary independent increments, and the mean and variance of the X_t,

$$E(X_t) = E(N_t)E(Y)$$
$$V(X_t) = E(N_t)E(Y^2). \tag{4}$$

See Parzen (1962) for some details about Poisson processes.

3. Methods for Estimating Change Point

The Maximum likelihood method and the Bayesian method are basic methods in statistical change point analyses for point estimation and hypothesis testing.

3.1 Change point estimation with the maximum likelihood method (frequentist method)

Let $X_1, X_2, \ldots, X_n$ be a sequence of the random variables such that X_i $(i = 1, \ldots, v)$ has probability density function $f(x, \theta_1)$ and X_i $(i = v + 1, \ldots, n)$ has probability density function $f(x, \theta_2)$ where the change point v is unknown discrete parameter and the parameters θ_1, θ_2 can be assumed to be either known or unknown. The single change point model in the sequence of the random variables is written:

$$X_1, X_2, \ldots, X_v \sim f(x, \theta_1)$$
$$X_{v+1}, X_{v+2}, \ldots, X_n \sim f(x, \theta_2). \tag{5}$$

Under model Eq.(5), for the observed values of the sequence of the random variables, the likelihood function is

$$L(\theta_1, \theta_2, v \mid x_1, \ldots, x_n) = L(\theta_1, \theta_2, v) = \prod_{i=1}^{v} f(x_i, \theta_1) \prod_{i=v+1}^{n} f(x_i, \theta_2) \tag{6}$$

and the logarithm of the likelihood function is:

$$\ell n L(\theta_1, \theta_2, v) = \sum_{i=1}^{v} \ell n f(x_i, \theta_1) + \sum_{i=v+1}^{n} \ell n f(x_i, \theta_2) \tag{7}$$

After adding the expression $\pm \sum_{i=1}^{v} \ell n f(x_i, \theta_2)$ in Eq.(7) then Eq.(7) can be re-written as follows:

$$\ell n L(\theta_1, \theta_2, v) \propto \sum_{i=1}^{v} \left[\ell n f(x_i, \theta_1) - \ell n f(x_i, \theta_2) \right] \tag{8}$$

When the parameters, θ_1, θ_2 are unknown, for any fixed values of change point v, the maximum likelihood estimator of the parameters θ_1, θ_2 are found to be the derivative of Eq.(8) θ_1 and θ_2 respectively:

$$\frac{\partial \ell nL(\theta_1, \theta_2; v)}{\partial \theta_1} = 0 , \tag{9}$$

$$\frac{\partial \ell nL(\theta_1, \theta_2; v)}{\partial \theta_2} = 0 . \tag{10}$$

After solving the equations system given in Eq.(9) and (10), the maximum likelihood estimator of the parameters θ_1, θ_2, $\hat{\theta}_1$, $\hat{\theta}_2$ are obtained. The maximum likelihood estimate of the change point v is:

$$\hat{v} = \arg \max_{k=1,\ldots n-1} \sum_{i=1}^{k} \left[\ell nf(x_i, \hat{\theta}_1) - \ell nf(x_i, \hat{\theta}_2) \right] \tag{11}$$

3.2 Change point estimation with the Bayesian method

The Bayesian method differs from the frequentist method in that each parameter is assumed to be a random variable and each one has a probability function called prior distribution. The estimate of the unknown parameter is obtained by deriving a posterior distribution on the basis of the prior distributions and the likelihood function. The posterior distribution is obtained:

Posterior $\propto$ likelihood $\times$ prior.

Under change point model given in Eq.(5), let $p_0(\theta_1, \theta_2 | v)$ be the joint prior distribution of parameters θ_1, θ_2 and let $p_0(v)$ be the prior distribution of change point v. The likelihood function is given by Eq.(6) and so the joint posterior distribution is written as follows:

$$p_1(\theta_1, \theta_2, v) \propto L(\theta_1, \theta_2, v) p_0(\theta_1, \theta_2 | v) p_0(v) . \tag{12}$$

Integrate Eq.(12) with respect to the parameters, the marginal posterior distribution of change point is proportional to

$$p_1(v) \propto \int_{\theta_2} \int_{\theta_1} L(\theta_1, \theta_2, v) p_0(\theta_1, \theta_2 | v) p_0(v) d\theta_1 d\theta_2 \tag{13}$$

and the Bayesian estimate $\hat{v}$ of the change point v is found by maximizing the marginal posterior distribution given by Eq.(13). Assuming uniform priors, the joint posterior mode, which gives the maximum likelihood estimates, is at $\hat{v}$, $\hat{\theta}_1$, $\hat{\theta}_2$ (Smith, 1975).

4. Detecting change point in worldwide earthquake data and findings

This study investigates whether there is a change point in worldwide earthquake activities such as the number of earthquake occurrences and their magnitude. At first, assuming that the occurrences of the earthquakes follow the homogeneous Poisson process, a change point is investigated in the occurrence rate of the earthquake in unit time; secondly, the magnitudes of the earthquake are assumed to be normally distributed random variables; a change point is investigated in the mean of the normally distributed random variables describing the earthquake magnitudes.

To detect a change point in earthquakes, the relevant data is taken from the website of the U.S. Geological Survey (2011), consisting of 819 earthquakes worldwide of magnitude 4.0 or above covering the period from 3-March-1901 until 11-March-2011. It is assumed that the earthquake occurrences follow the homogeneous Poisson process. The earthquakes are observed at times $S_1, S_2, ..., S_{818}$ in the continuous time interval $(0,T]$. For the earthquakes data, the starting point S_0 is in 3-March-1901 and the end point $S_{818}=T$ is in 11-March-2011. $T_1, T_2, ..., T_{818}$ are exponentially distributed random variables as given in Eq.(2). Under the assumption that τ is equal to the occurrence time of an event, the likelihood function can be written under the change point model in the sequence of exponentially distributed random variables:

$$L(\lambda_1, \lambda_2, \tau) = \lambda_1^{N_\tau} e^{-\lambda_1 \tau} \lambda_2^{n-N_\tau} e^{-\lambda_2(T-\tau)} . \qquad (14)$$

Where τ shows change point and is a continuous parameter defined time interval $(0,T]$, λ_2 is the occurrence rate in the time interval $(\tau, T]$ and N_τ is the number of events that occurred in the time interval $(0,\tau]$ and T is the sum of the time interval between the earthquake occurrences, that is $T = \sum_{i=1}^{818} t_i$ (Raftery & Akman, 1986; Akman & Raftery, 1986).

The logarithm of the likelihood function is

$$\ell nL(\lambda_1, \lambda_2, \tau) = N_\tau \ell n(\lambda_1/\lambda_2) - \tau(\lambda_1 - \lambda_2) + n\ell n\lambda_2 - \lambda_2 T \qquad (15)$$

After the derivation of the log likelihood function given by Eq.(15) for λ_1 and λ_2, the maximum likelihood estimations of λ_1 and λ_2 are obtained respectively:

$$\hat{\lambda}_1 = \frac{N_\tau}{\tau}, \quad \hat{\lambda}_2 = \frac{n-N_\tau}{T-\tau} \qquad (16)$$

For the possible values of τ, $s_1, s_2, ..., s_n$, and the variable N_τ takes the values $1, 2, ..., n$, and so the estimations $(\hat{\lambda}_{11}, \hat{\lambda}_{21}), (\hat{\lambda}_{12}, \hat{\lambda}_{22}), ..., (\hat{\lambda}_{1n}, \hat{\lambda}_{2n})$ are calculated from Eq.(16) and then these estimations are substituted into Eq.(15). One of $(\hat{\lambda}_{1i}, \hat{\lambda}_{2i}, s_i)$, $i = 1, 2, ..., n$, gives the maximum of this function, say $\hat{\lambda}_1, \hat{\lambda}_2, \hat{\tau}$. Hence $\hat{\lambda}_1, \hat{\lambda}_2$ and $\hat{\tau}$ are the maximum likelihood estimation for λ_1, λ_2 and τ.

The scatter plot of the time intervals between the earthquakes is shown in Figure 1. From Figure 1, it is clearly seen that there is a change (or at least one change) in the occurrence rate of the earthquakes.

When we assume a single change point in the data and the method explained above is applied this data and the results are given in Table 1.

As is shown in Table 1, the occurrence rate of the earthquake in unit time (in day or in year) increased considerably after February 3, 2002, approximately ten times of the occurrence rate of the earthquakes in unit time before February 3, 2002.

From the log likelihood function of the change point in Figure 2, it can be seen that there are at least two change points in the data. There many methods are used to detect multiple change points in the data. One of the basic methods used for this purpose is binary segmentation procedure (Chen & Gupta, 1997; Yang & Kuo, 2001). In the binary segmentation procedure, the data is divided into two homogeneous groups according to the estimated change point, and a change point is searched in each subdivided data until there is no change in the subdivided data. The results are given in Figure 3.

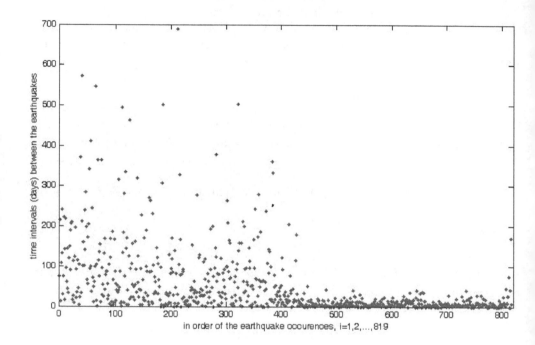

Fig. 1. The time intervals (days) between the earthquakes according to the earthquake occurrences.

n = 818, T= 40185 days		
Before the change point $\hat{\lambda}_1$	Estimated change point $\hat{\tau}$ $N_{\hat{\tau}}$ =429	After the change point $\hat{\lambda}_2$
0.0116380 (in days) 4.1896804 (in years)	03-Feb-2002	0.1170628 (in days) 42.1426080 (in years)

Table 1. Estimated parameters for the earthquake data.

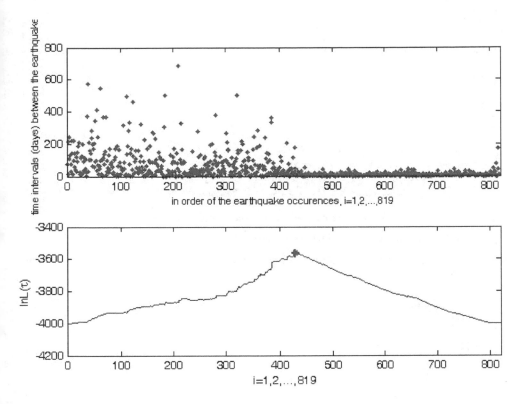

Fig. 2. The maximum point of log likelihood function given in Eq.(16) is shown as a red star and the corresponding change point estimate is shown with a red circle.

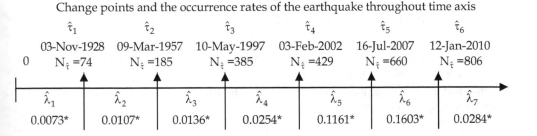

Fig. 3. The maximum likelihood estimations of parameters for multiple change points in the earthquake data (*in days).

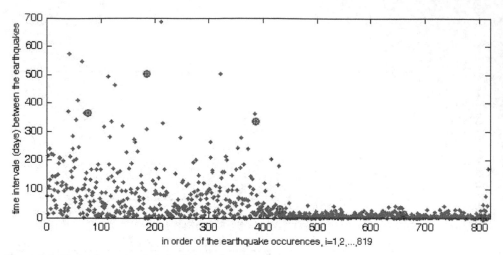

Fig. 4. Multiple change points in the earthquake data.

Estimated change points are shown on the scatter plot of the time intervals between the earthquakes.

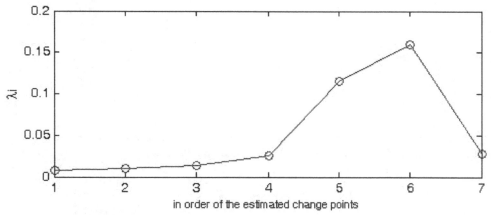

Fig. 5. The estimated occurrence rate of earthquakes in order of change points.

As can be seen in Figure 4 and Figure 5, the occurrence rate of the earthquakes increases slowly up until 03-February-2002, goes up sharply between 03- February-2002 and 12-January-2010, and tends to fall thereafter.

When using Bayesian method, the prior distributions of the parameters λ_1, λ_2 and τ, are taken to be respectively,

$$p_0(\lambda_1, \lambda_2 \mid \tau) = \frac{1}{\lambda_1 \lambda_2}, \qquad \lambda_1, \lambda_2 > 0$$

$$p_0(\tau) = \frac{1}{T}, \qquad 0 < \tau < T$$

(17)

where the prior distributions are called uninformative priors. The joint posterior distribution of the parameters is can be written as:

$$p_1(\lambda_1,\lambda_2,\tau) \infty L(\lambda_1,\lambda_2,\tau)p_0(\lambda_1,\lambda_2 \mid \tau)p_0(\tau)$$
$$\infty \lambda_1^{N_\tau}e^{-\lambda_1\tau}\lambda_2^{n-N_\tau}e^{-\lambda_2(T-\tau)}\frac{1}{\lambda_1\lambda_2 T} \tag{18}$$

With integrate Eq. (18) with respect to the parameters λ_1,λ_2, the marginal posterior distribution of change point is proportional to

$$p_1(\tau) \propto \frac{\Gamma(N_\tau)\Gamma(n-N_\tau)}{\tau^{N_\tau}(T-\tau)^{n-N_\tau}}\left(\frac{1}{T}\right) \tag{19}$$

where $\Gamma(.)$ is gamma function. The marginal posterior distributions of λ_1 and λ_2 are not obtained analytically because of discontinuity in τ and the close form of the marginal posterior distributions of λ_1 and λ_2 are respectively proportional to

$$p_1(\lambda_1) \propto \frac{1}{T}\int_0^T \lambda_1^{N_\tau}e^{-\lambda_1\tau}\frac{\Gamma(n-N_\tau)}{(T-\tau)^{n-N_\tau}}d\tau,$$

$$p_1(\lambda_2) \propto \frac{1}{T}\int_0^T \lambda_2^{n-N_\tau}e^{-\lambda_2(T-\tau)}\frac{\Gamma(N_\tau)}{\tau^{N_\tau}}d\tau$$

where, for the possible values of τ, $s_1,s_2,\ldots,s_n$, and the variable N_τ takes the values $1,2,\ldots,n$.

Form Eq.(19), the Bayesian estimate $\hat\tau$ of the change point in the earthquake data is found to be the date 03-February-2002 2002 which corresponds to the posterior mode. The posterior distribution of the change point is given in Figure 6.

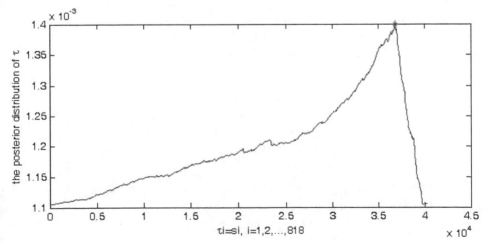

Fig. 6. The posterior distribution of change point and the posterior mode.

The Bayesian estimates of the parameters λ_1 and λ_2 would be the same as the maximum likelihood estimates because of uninformative prior distributions.

For multiple change points, the binary Bayesian segmentation procedure can be easily used. The procedure is employed as the mode of the posterior distribution of change point decreases considerably until the one before the posterior mode is found. The results are given in Figure 7.

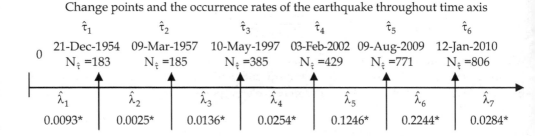

Change points and the occurrence rates of the earthquake throughout time axis

$\hat{\tau}_1$	$\hat{\tau}_2$	$\hat{\tau}_3$	$\hat{\tau}_4$	$\hat{\tau}_5$	$\hat{\tau}_6$
21-Dec-1954	09-Mar-1957	10-May-1997	03-Feb-2002	09-Aug-2009	12-Jan-2010
$N_{\hat{\tau}}$ =183	$N_{\hat{\tau}}$ =185	$N_{\hat{\tau}}$ =385	$N_{\hat{\tau}}$ =429	$N_{\hat{\tau}}$ =771	$N_{\hat{\tau}}$ =806

$\hat{\lambda}_1$	$\hat{\lambda}_2$	$\hat{\lambda}_3$	$\hat{\lambda}_4$	$\hat{\lambda}_5$	$\hat{\lambda}_6$	$\hat{\lambda}_7$
0.0093*	0.0025*	0.0136*	0.0254*	0.1246*	0.2244*	0.0284*

Fig. 7. The Bayesian estimations of parameters for multiple change points in the earthquake data (*in days).

When comparing the maximum likelihood estimates and the Bayesian estimates of the change points, we can see that the change points corresponding to dates such as 9-March-1957, 10-May-1997, 03-February-2002 and 12-January-2010 overlap. These dates are investigated in depth from many aspects, including overall world temperature, other disasters, and astronomical events.

When we look at the magnitudes of the earthquakes in the data, those of magnitude 6 to 6.9 number 291 (35.57%) and those of magnitude 7 to 7.9 number 258 (31.54%) (Table 2). The histogram of the magnitudes is given in Figure 6. Furthermore, using the Kolmogorov-Smirnov test, the distribution of the magnitudes is found to be almost normal (p=0.000).

Magnitude	4-4.9	5-5.9	6-6.9	7-7.9	8-8.9	≥9
The number of earthquakes	54	93	291	258	40	82
%	0.0660	0.1137	0.3557	0.3154	0.0489	0.1002

Table 2. The number of earthquakes with respect to magnitudes.

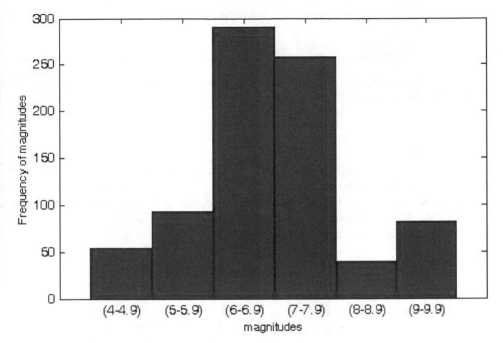

Fig. 8. Histogram of the earthquake magnitudes.

The sequence of the earthquake magnitudes $Y_1, Y_2, \ldots, Y_n$ can be assumed to be normally distributed random variables. The change point model in the mean of the normally distributed random sequences is written,

$$
\begin{aligned}
Y_1, Y_2, \ldots, Y_v &\sim N(\mu_1, \sigma^2) \\
Y_{v+1}, Y_{v+2}, \ldots, Y_n &\sim N(\mu_2, \sigma^2)
\end{aligned}
\tag{20}
$$

where v is an unknown change point in the mean of the sequence of the normally distributed random variables. To investigate whether there is a change point in the earthquake magnitudes, under the change point model, the likelihood function of the observed values, $y_1, y_2, \ldots, y_n$,

$$
\begin{aligned}
L(\theta_1, \theta_2, v) &= \prod_{i=1}^{v} f(y_i, \mu_1, \sigma^2) \prod_{i=v+1}^{n} f(y_i, \mu_2, \sigma^2) \\
&\propto \frac{1}{(\sqrt{2\pi\sigma^2})^n} \exp\left\{ -\frac{1}{2\sigma^2} \left(\sum_{i=1}^{v} (y_i - \mu_1)^2 - \sum_{i=v+1}^{n} (y_i - \mu_2)^2 \right) \right\}
\end{aligned}
\tag{21}
$$

and the logarithm of the likelihood function in Eq. (21) is:

$$
\ell n L(\theta_1, \theta_2, v) \propto -\frac{1}{2\sigma^2} \left(\sum_{i=1}^{v} (y_i - \mu_1)^2 - \sum_{i=1}^{v} (y_i - \mu_2)^2 \right)
\tag{22}
$$

For any given value of v, the maximum estimates of unknown parameters μ_1, μ_2, are

$\bar{y}_1 = \dfrac{\sum_{i=1}^{v} y_i}{v}$ and $\bar{y}_2 = \dfrac{\sum_{i=v+1}^{n} y_i}{n-v}$ respectively. The estimates are substituted into Eq.(22), and

with the maximized the log likelihood function given by Eq.(22), we can obtain the maximum likelihood estimate of the change point,

$$\hat{v} = \arg \max_{k=1,\dots n-1}\left[\frac{1}{2\sigma^2}\left(v\bar{y}_1^{-2} - (n-v)\bar{y}_2^{-2}\right)\right] \tag{23}$$

From Eq. (23), the change point is estimated to be the date 28-February-2011 corresponding to v=817. The estimated change point is close to end of the sequence such that it refers to no change point in the sequence. We can model both the number of the earthquake occurrences and the magnitudes of the earthquakes with compound Poisson process. Let $\{X_t, t \geq 0\}$ be a compound Poisson process as defined in Subsection 2.1

$$X_t = \sum_{i=1}^{N_t} Y_i$$

where N_t is the number of earthquakes in time interval $(0,t]$, which is a Poisson distributed random variable with parameter λt, and $\{Y_i, i = 1,2,\dots\}$ are the magnitudes of the earthquakes, which are normally distributed random variables with parameters μ_i and σ^2. The mean and variance of the magnitudes of the earthquakes according to the estimated change points dates, (9-March-1957, 10-May-1997, 03-February-2002 and 12-January-2010) are given in Table 3.

Estimated Change points	Up to 9-Mar-1957	Between 9-Mar-1957 and 10-May-1997	Between 10-May-1997 and 03-Feb-2002	Between 03-Feb-2002 and 12-Jan-2010	After 12-Jan-2010
Occurrence rate of the earthquakes	0.0090*	0.0136*	0.0254*	0.1300*	0.0284*
Mean of the magnitudes	6.9622	6.7250	7.0795	6.4562	7.1167
Variance of the magnitudes	0.9305	0.8821	0.5579	0.9845	1.2706

Table 3. Means and variances of the magnitudes with respect to estimated change point (*in days).

Using Table 3, we compute the recurrence periods for certain earthquakes of magnitude y, corresponding to the change point, 12-January-2010. The recurrence period of an earthquake with magnitude y is computed by

Reccurence period=1/(Expected number of earthquakes).

The expected number of earthquakes of magnitude y is computed by the multiplying the probability of occurrence class of magnitude with the expected number of earthquakes over a certain period of time.

Class Magnitude (y)	$F_Y(L)$ Expected	$F_Y(U)$ Expected	$P(L \leq y \leq U)$ Expected	Expected number of earthquakes (year)	Average recurrence period (year)
4-4.9	0.0028	0.0246	0.0218	0.2257	4.4306
5-5.9	0.0302	0.1402	0.1100	1.1404	0.8769
6-6.9	0.1609	0.4238	0.2629	2.7248	0.3670
7-7.9	0.4588	0.7565	0.2977	3.0856	0.3241
8-8.9	0.7834	0.9432	0.1598	1.6566	0.6037
≥9	0.9526	1.0000	0.0474	0.4912	2.0360

Table 4. The estimation of recurrence periods for certain earthquakes after the change point 12-January-2010.

Time	Expected number of earthquakes	Expected total magnitudes of earthquakes
30 days	0.8520	6.0634
3 months	2.5560	18.1902
6 months	5.1120	36.3805
1 years	10.3660	73.7717
2 years	20.7320	147.5434
3 years	31.0980	221.3151
4 years	41.4640	295.0868
5 years	51.8300	368.8585
10 years	103.6600	737.7171
20 years	207.3200	1475.4342

Table 5. Expected number of earthquakes and the expected total magnitudes of earthquakes with respect to the change point 12-January-2010.

Number of earth.	30 days 11-Feb-10	3 months 12-Apr-10	6 months 11-Jul-10	1 year 12-Jan-11	2 years 11-Jan-12	3 years 11-Jan-13	4 years 11-Jan-14	5 years 11-Jan-15	10 years 11-Jan-20	20 years 11-Jan-30
0	0.426560956	0.077614579	0.006024023	3.1485E-05	9.91304E-10	3.12112E-14	0	0	0	0
1	0.363429935	0.198382863	0.030794805	0.000326373	2.05517E-08	9.70605E-13	0	0	0	0
2	0.154821152	0.253533299	0.078711521	0.001691593	2.13039E-07	1.50919E-11	0	0	0	0
3	0.043969207	0.216010371	0.134124431	0.005845017	1.47224E-06	1.56443E-10	1.16755E-14	0	0	0
4	0.009365441	0.138030627	0.171411023	0.015147361	7.63063E-06	1.21627E-09	1.21028E-13	0	0	0
5	0.001595871	0.070561256	0.17525063	0.031403509	3.16396E-05	7.56469E-09	1.00366E-12	0	0	0
6	0.000226614	0.030059095	0.149313537	0.054254796	0.000109326	3.92078E-08	6.93599E-12	0	0	0
7	2.75821E-05	0.010975864	0.109041543	0.080343602	0.000323791	1.74183E-07	4.10848E-11	0	0	0
8	2.9375E-06	0.003506789	0.069677546	0.104105222	0.000839104	6.77095E-07	2.12943E-10	3.99619E-14	0	0
9	2.78083E-07	0.000995928	0.039576846	0.119906081	0.001932923	2.33959E-06	9.8105E-10	2.30136E-13	0	0
10	2.36927E-08	0.000254559	0.020231684	0.124294643	0.004007336	7.27565E-06	4.06783E-09	1.1928E-12	0	0
11	1.83511E-09	5.91503E-05	0.009402215	0.117130752	0.007552736	2.05689E-05	1.53335E-08	5.62024E-12	0	0
12	1.30292E-10	1.2599E-05	0.004005344	0.101181448	0.01304861	5.33044E-05	5.29823E-08	2.42747E-11	0	0
13	8.53917E-12	2.47716E-06	0.001575024	0.08068053	0.020809521	0.000127512	1.68989E-07	9.67815E-11	3.26026E-14	0
14	5.19669E-13	4.52259E-07	0.000575109	0.05973817	0.030815928	0.000283241	5.00498E-07	3.58299E-10	1.44839E-13	0
15	2.95172E-14	7.70649E-08	0.000195997	0.041283058	0.042591721	0.000587215	1.38351E-06	1.23804E-09	6.00562E-13	0
16	0	1.23111E-08	6.26211E-05	0.026746261	0.055188223	0.001141327	3.58536E-06	4.01048E-09	2.33453E-12	0
17	0	1.85101E-09	1.88305E-05	0.016308926	0.067303661	0.002087822	8.74491E-06	1.22273E-08	8.5411E-12	0
18	0	2.62844E-10	5.34787E-06	0.009392129	0.077518861	0.003607061	2.01444E-05	3.52077E-08	2.95123E-11	0
19	0	3.53594E-11	1.43886E-06	0.005124148	0.084585317	0.005903809	4.39614E-05	9.60429E-08	9.66079E-11	0
20	0	4.51893E-12	3.67772E-07	0.002655846	0.08768114	0.009179833	9.11408E-05	2.48895E-07	3.00431E-10	0
25	0	0	2.01378E-10	4.98583E-05	0.052673373	0.041877043	0.001752055	1.46016E-05	4.38567E-08	0
30	0	0	4.11104E-14	3.48963E-07	0.011797303	0.071223597	0.012557101	0.000319369	2.3869E-06	0
35	0	0	0	1.07217E-09	0.001159888	0.05317575	0.039506887	0.003066385	5.70261E-05	0
40	0	0	0	1.6252E-12	5.62613E-05	0.019586801	0.061322008	0.014525139	0.000672161	4.94125E-13
150	0	0	0	0	0	0	0	0	0	3.67978E-06

Table 6. Probability of earthquake occurrences after the change point 12-January-2010.

5. Conclusion

Predicting earthquakes is of crucial importance these days. Many researchers have studied earlier attempts at earthquake detection. The change point analysis is used in both backward (off-line) and forward (on-line) statistical research. In this study, it is used with the backward approach in the worldwide earthquake data. The change points found in the worldwide earthquake data are useful in making reliable inferences and interpreting the results for further research. Each date found as the change point in the earthquake data should be carefully investigated with respect to other geographical, ecological, and geological events or structures in the world.

6. References

Akman V. E. & Raftery A. E. (1986). Asymptotic inference for a change-point Poisson process, *The Annals of Statistics*, Vol.14, No.4, pp. 1583–1590.

Aktaş, S.; Konşuk, H. & Yiğiter, A. (2009). Estimation of change point and compound Poisson process parameters for the earthquake data in Turkey, *Environmetrics*, Vol. 20, No.4, pp. 416-427

Amorese, D. (2007). Applying a Change-Point Detection Method on Frequency-magnitude Distributions, *Bulletin of the Seismological of America*, Vol.97, No.5, pp.1742-1749

Boudjellaba, H.; MacGibbon, B. & Sawyer, P. (2001). On exact inference for change in a Poisson Sequence. *Communications in Statistics A: Theory and Methods*, Vol.30, No.3, pp. 407–434

Chen J. & Gupta A. K. (1997). Testing and locating variance change points with application to stock prices, Journal of the American Statistical Association: JASA, Vol.92, No. 438, pp. 739-747

Fotopoulos S. B. & Jandhyala V. K. (2001). Maximum likelihood estimation of a change-point for exponentially distributed random variables, *Statistics & Probability Letters*, 51: 423–429

Kagan, Y. Y. (2010). Statistical Distributions of Earthquake Numbers: Consequence of Branching Process, *Geophysical Journal International*, Vol.180, No.3, pp.1313-1328

Leonard T. & Papasouliotis, O. (2001). A Poisson Model for Identifying Characteristic Size Effects in Frequency Data: Application to Frequency-size Distributions for Global Earthquakes, "Starquakes", and Fault Lengths, *Journal of Geophysical Research*, Vol.106, No.B7, pp.13,413-13,484

Lindsey H. (June 2007). 'Natural' catastrophes increasing worldwide, 14.06.2007, Available from: http://www.wnd.com/news/article.asp?ARTICLE_ID=40584/

Mandeville M. W. (June 2007). "Eight Charts Which Prove That Chandler's Wobble Causes Earthquakes, Volcanism, El Nino and Global Warming", 14.06.2007, Available from: http://www.michaelmandeville.com/polarmotion/spinaxis/vortex_correlations2. htm

Parzen, E. (1962). *Stochastic Processes*, Holden-Day, ISBN 0-8162-6664-6, San Francisco, USA

Raftery A. E. & Akman V. E. (1986). Bayesian analysis of a Poisson process with a change point, *Biometrika*, Vol.73, pp.85–89

Rotondi R. & Garavaglia, E. (2002). Statistical Analysis of the Completeness of a Seismic Catalogue, *Natural Hazards*, Vol.25, No. 3, pp. 245-258

Rydelek, P. A. & Sacks I. S. (1989). Testing the completeness of earthquake catalogues and the hypothesis of self-similarity, *Nature*, Vol. 337, pp.251–253

Schorlemmer, D. & Woessner, J. (2008). Probability of Detecting an Earthquake, *Bulletin of the Seismological of America*, Vol.98, No.5, pp. 2103-2117

Smith, A. F. M. (1975).A Bayesian approach to inference about a change point in a sequence of random variables, *Biometrika*,Vol.62, pp. 407-416

U.S. Geological Survey web page 16. 03. 2011, Available from:
    http://earthquake.usgs.gov/regional/world/historical.php

Yang, T. Y. & Kuo, L. (2001). Bayesian binary segmentation procedure for a Poisson process with multiple change points, *Journal of Computational and Graphical Statistics*, Vol. 10, pp. 772–785

Yao, Q. (1993). Tests for change-points with epidemic alternatives. *Biometrika*, Vol.80, pp. 179-191

Yigiter, A. & İnal, C. (2010). Estimation of change points in a homogeneous Poisson process with an application to earthquake data, *Pakistan Journal of Statistics*, Vol. 26, No. 3, pp. 525-538

Part 2

Studies on Earthquake Precursors and Forecasting

3

Current State of Art in Earthquake Prediction, Typical Precursors and Experience in Earthquake Forecasting at Sakhalin Island and Surrounding Areas

I.N. Tikhonov[1] and M.V. Rodkin[1,2]
[1]Institute of Marine Geology and Geophysics FEB RAS, Yuzhno-Sakhalinsk
[2]International Institute of Earthquake Prediction Theory and Mathematical
Geophysics RAS, Moscow
Russia

1. Introduction

Despite of over a century of scientific effort, the understanding in earthquake forecasting remains immature. Moreover, even the theoretical possibility of earthquake forecasting is debatable. Especially problematic is a possibility of an effective short- and intermediate-term earthquake forecasting. The aim of this paper is to present the new evidence in support of possibility of the short- and intermediate-term earthquake forecasting. This possibility is shown through the discussion of seismic regime in the generalized vicinity of strong earthquake and through the description of an experience in an earthquake forecasting in the case of the Sakhalin Island and the surrounding areas.

USGS/NEIC catalog and Harvard seismic moments catalog are used to construct the generalized space–time vicinity of strong (M7+) earthquake to reveal the robust typical long-, intermediate, and short-term precursor anomalies. The very essential increase in available information resulted from this procedure gives possibility to detail the character of precursors of strong earthquake. The typical parameters of the fore- and aftershock cascades were detailed. A few other revealed precursory anomalies indicate the development of softening in the source area of a strong earthquake. The set of the precursory anomalies indicates the approaching of a strong event quite definitely. Thus one can conclude that the effective short- and intermediate-term earthquake forecasting appears to be possible in the case of essential increase of volume of statistical information available for the forecasting.

The current state of art in the earthquake forecasting is illustrated by the case of experience in the earthquake forecasting for the Sakhalin Island and the surrounding areas performed in the Institute of Marine Geology and Geophysics of the Far East Branch of the Russian Academy of Science, Yuzhno-Sakhalinsk, Russia. Four examples of successful prognosis (three of them performed in a real time), and one false alarm took place. Thus, despite the evident deficient in available information the results of forecasting appear to be encouraging enough. In any case they are much better than they could be in the case if the seismic roulette model would be valid.

In the early 1980s a few examples of successful earthquake prognosis were known, and the final successes in decision of the problem of earthquake prognosis seemed to be close. But the substantial increase in a number of different sensors used in earthquake monitoring, and the corresponding increase in available information didn't improve the quality of prognosis. The situation was discussed widely in the 90s, and the dominant opinion elaborated by the world scientific community was quite pessimistic. An earthquake generating system was found to be very unstable. A minor change in parameters of such systems can significantly change their evolution; as a result an effective prognosis of behavior of such systems is impossible. Thus an earthquake prognosis was declared to be impossible (Geller, 1997; Geller et al., 1997; Kagan, 1997; and references herein). Despite of this dominating opinion a few groups of researchers have continued their investigations in earthquake forecasting. First of all the effectiveness of the suggested earlier algorithms of strong earthquake prediction was tested in real time. The results of the use of the M8 and Mendocino Scenario algorithms suggested earlier in (Keilis-Borok & Kossobokov, 1986, 1990; Kossobokov, 1986) were examined during more than twenty years. It was shown that the results of prognosis were significantly better than it could be in case of a seismic roulette procedure (Shebalin, 2006; Kossobokov, 2005). However neither these algorithms nor the other ones tested at shorter time intervals (Sobolev et al., 1999; Papazachos, 2005; Zavyalov, 2006; and others) showed results quite suitable for practical use. There were substantial probabilities to miss an earthquake or declare false alarm.

Low efficiency of earthquake prediction is connected to extremely irregular character of seismic regime. Due to the high level of irregularity of seismic regime parameters of earthquake precursors are vague, and even the very existence of precursor phenomena remains debatable. As a result in the absence of well known precursors any algorithm of forecasting based on the use of these precursors could hardly be very effective.

Thus verification of used precursor phenomena is an urgent problem. A precursory process and occurrence of large earthquake is commonly treated as an example of critical phenomenon (Akimoto & Aizawa, 2006; Bowman et al., 1998; Keilis-Borok & Soloviev, 2003; Malamud et al., 2005; Nonlinear ..., 2002; Sornette, 2000; etc.). Many of the precursors used currently, such as development of foreshock cascade, an increase in correlation length, and an abnormal clustering of earthquakes, are expected to occur in critical processes. Moreover, some of these precursors came in the use because the process of strong earthquake occurrence is treated in terms of the critical phenomenon model. In this situation a natural question may arise: to what extent are such model processes really typical of scenarios of occurrence of large earthquake? Romashkova and Kossobokov (2001) have considered the evolution of foreshock and aftershock activity in the vicinities of eleven strong earthquakes occurring from 1985 to 2000. This examination has not supported the universality of power-law growth in foreshock activity toward the moment of a large earthquake. It also turned out that the aftershock sequences in a number of cases differ significantly from the Omori law. As a result it was hypothesized (Romashkova & Kossobokov, 2001; Kossobokov, 2005) that scenarios of aftershock sequences deviating from the Omori law can exist.

It seems natural to ask whether the observed deviations of the seismic process from the theoretically expected universal scenario have a stochastic nature or different scenarios can be put into effect in different foreshock and aftershock sequences. The answer to that question can be obtained by investigation of mean features inherent to vicinities of a large number of strong earthquakes. A strong earthquake vicinity is understood here as a space-time domain where evolution of seismicity is influenced by occurrence of a given strong earthquake. Using the approach presented in (Rodkin, 2008) we have constructed the mean generalized space–time vicinity of a large number of strong earthquakes and examined the mean anomalies inherent to this vicinity.

2. Construction of generalized vicinity of strong earthquake

We have used the Harvard worldwide seismic moment catalog for 1976-2005, and the USGS/NEIC catalog for 1968-2007. In both cases only shallow earthquakes with depth H < 70 km were examined. Two subsets of data can be used, first one includes all earthquakes from the catalog and the second includes stronger earthquakes that are only completely reported. Below we present the results from processing of the Harvard catalog using the first subset of data (all reported events) and the results for the USGS/NEIC catalog using only completely reported events. In the latter case the events with magnitude M ≥ 4.7 were used, a total number of events was 97615. A similar cutoff for the Harvard catalog would reduce the available data too much to get statistically robust results.

Both used data sets were searched for events falling into the space–time domains surrounding the source zones of large (M7+) earthquakes, with due account for the seismic moment in the Harvard catalog and the maximum magnitude for the USGS/NEIC catalog. A generalized vicinity of large earthquake is understood as a set of events falling into the zone of influence of any of these strong earthquakes. The zones of influence were defined as following, see also (Rodkin, 2008) for the details. Spatial dimensions of the zones of influence for earthquakes of different magnitudes were calculated from the approximate relationship (Sobolev & Ponomarev, 2003) between typical source size L and earthquake magnitude M:

$$L \text{ (km)} = 10^{0.5M - 1.9}. \tag{1}$$

In the examination below the earthquakes located at distances within $7{\times}L$ from the epicenter of the given strong earthquake were taken into account.

For constructing a time vicinity of strong earthquake we used the conclusion that duration of a failure cycle weakly depends on earthquake magnitude (Smirnov, 2003). Hence the simple epoch superposition method can be used for comparing the time vicinities of earthquakes with close magnitudes. At the figures below all earthquakes located in the area $7{\times}L$ of the corresponding strong event were taken into account. This choice allows the most complete use of available data. Negative consequences of this choice are a lower statistical significance at the edges of the time interval because of shortage of data there, and a false effect of a systematic growth of a number of earthquakes towards the centre of the used time interval. However these errors can be taken into account, so they do not distort the results.

The generalized vicinity of large earthquake which was constructed contained more than 60000 earthquakes for the Harvard catalog and more than 300000 earthquakes for the USGS/NEIC catalog. Such a big number of events resulted from the fact that one and the same earthquake can belong to the space–time vicinities of different strong earthquakes.

Such an increase in a number of events has considerably enhanced the possibility of statistical examination.

Time and space position of each earthquake falling into the generalized vicinity of large earthquake is characterized by the time shift from the origin time of the corresponding strong earthquake and by the distance from the epicenter of this main event (norm to the source size of this main event). Both catalogs (USGS/NEIC and Harvard) were used for examination of the relative space-time density of earthquakes. The Harvard catalog was used in this paper mostly for the verification of results, which were obtained from examination of the USGS/NEIC catalog.

3. Regularities in rate of fore- and aftershock cascades

The most well known feature of seismic behavior occurring in the vicinities of large earthquakes is the existence of aftershock and foreshock power-law cascades. Figures 1a and 1b show the foreshock and aftershock sequences in the generalized vicinity of strong earthquake, which were obtained from USGS/NEIC data (similar results were obtained from examination of the Harvard catalog). The earthquakes rate is presented by time density of group of earthquakes consisting of subsequent 50 events taken with step 25 events (rate of events is given in n/day for convenience).

As can be seen in Fig. 1, the evolution of foreshocks and aftershock sequences well correlates with a power law. The Omori law (Utsu et al., 1995; Sobolev, 2003) is known to be a good fit to the aftershock rate:

$$n \sim 1/ \ (c + t)^{-p}, \tag{2}$$

where n is the rate of aftershock occurrences, t is the time interval after the main shock occurrence, c – parameter fitting the rate of earthquakes in the closest vicinity of the main shock, and p is the parameter of the Omori law. The Omori law (2) is a good fit for the interval until one hundred days or somewhat later after the main shock occurrence.

The foreshock cascade occurring before the main shock time can be described in a similar manner; in this case t is the time before the main shock origin, and c = 0. The foreshock cascade was found to be quite noticeable in the generalized vicinity 10-20 days before the main shock occurrence (Fig. 1).

Of special interest is the deviation of the aftershock rate from the power law during the first hours after the main shock occurrence. The deficit of earlier aftershocks described by parameter c in (2) is explained sometimes by difficulty in recording all of too numerous aftershocks occurring immediately after a large earthquake. However, this factor is hardly capable of providing a full explanation of the phenomenon (Lennartz et al., 2008; Shebalin, 2006). The deviation from the power law toward lower rates of events during a few first hours of the aftershock sequence can be seen clearly in Fig. 1b; the rate of aftershocks reaches the values obeying the power law only 2–3 hours after the strong earthquake. At that time the mean rate of earthquakes with M ≥ 4.7 occurring in the vicinity of a mean large earthquake (but not in the generalized vicinity of large earthquake) is a little above one event per hour. Such rate can not cause any problem in events recording. Thus, the effect of a lower rate of earlier aftershocks probably has a physical nature. This conclusion is similar with those presented in (Lennartz et al., 2008; Lindman et al., 2010; Shebalin, 2006).

Current State of Art in Earthquake Prediction, Typical Precursors and Experience in Earthquake Forecasting at Sakhalin
Island and Surrounding Areas

47

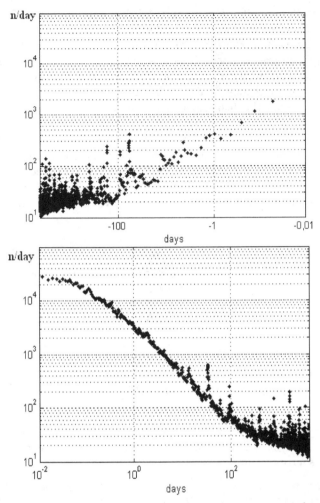

Fig. 1. The fore- (upper panel) and aftershock (lower panel) sequences in the generalized vicinity of strong earthquake, events flow is given in a number of events per day, zero time corresponds to the moment of occurrence of the generalized main shock.

Note that as it can be seen in Fig. 1, the mean duration of the foreshock process is significantly shorter than that of the aftershocks, and the rate of increase for foreshocks toward the moment of main shock occurrence is noticeably slower than the decay of the aftershock rate. From this it follows that the typical maximum rate of foreshock sequence is an order smaller than the maximum rate of aftershock process. This result is intimately related to the problem of predictability of large earthquakes; the prediction problem would have been solved already, if the rate of foreshocks would be equal to the rate of aftershocks. It seems important to note that upon closer examination the seismicity increase in the generalized vicinity of large earthquakes is not confined to the foreshock and aftershock cascades. Essentially weaker but quite noticeable increase above the mean rate level occurs in the time interval about ±100-300 days around the main shock date. The analysis of the Harvard seismic moment catalog gives similar results; however, these data testify for a broader area of seismicity increase, roughly within ±500 days around the main shock date. This type of long-term pre- and post-shock seismic activity agrees with the suggestion that the final time interval of strong earthquake preparation prolongs a few years.

We now characterize the changes in the seismicity increase as functions of the distance to the main shock epicenter. The distances are compared in units of the magnitude-dependent main shock source dimension L from equation (1). Fig. 2 shows the distance–time diagram of rate of a number of events in the vicinity of the main shock. The horizontal axis indicates the time (in days) from the main shock occurrence time; an analogue of longtime scale is used near the main shock occurrence moment. The vertical axis indicates the distance from the main shock epicenter in units of earthquake source size L. Events' rate is given in logarithmic scale, $\ln(n)$, where n is the number of events in a cell of the distance–time diagram.

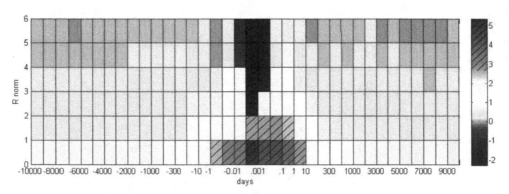

Fig. 2. Spatial-temporal change of number of earthquakes lg (density of events per day) in the generalized vicinity of strong earthquake, distance R_{norm} from the main shock epicenter is given in norm source size units.

As can be seen in Fig. 2, the rate of earthquakes in the vicinity of the main shock begins to increase one-three hundred days before the main shock, and this activity increase accelerates toward the moment of the main shock. The increase in seismic activity occurs at a distance of about three source sizes L from the main shock. This estimate of the radius of influence is in agreement with the size of the areas where predictive functions are usually estimated in earthquake prediction algorithms (Kossobokov, 2005; Shebalin, 2006). Seismic activity outside the zone of 3–4 earthquake source sizes decreases in the close time vicinity of strong earthquake. This feature can result from softening in the source of ongoing strong earthquake. Some other indications of strength decrease in the strong earthquake vicinity are presented in (Rodkin, 2008). In this case one can expect the strain rearrangement from the outer "rigid" region into the inner "soft" zone where the strong earthquake is about to occur. As a result of this rearrangement, the probability of earthquake occurrence in the outer "rigid" zone of the future rupture would become somewhat lower.

The decrease in b-value is known to be used as an indicator of an increase in probability of a strong earthquake occurrence (Shebalin, 2006; Zavyalov, 2006). Catalog USGC/NEIS was used to examine the change in b-value in the generalized vicinity of strong earthquake (the similar results were obtained from examination of the Harvard catalog). The maximum likelihood method was used for the b-values estimation (Utsu, 1965). By this method the b-value is calculated from

$$b = lg(e) / (M_{av} - M_c) \qquad (3)$$

where M_{av} is the average magnitude for each subset of data and M_c is the lower magnitude limit used in the analysis, here M_c=4.7. Discreteness of magnitude values because of

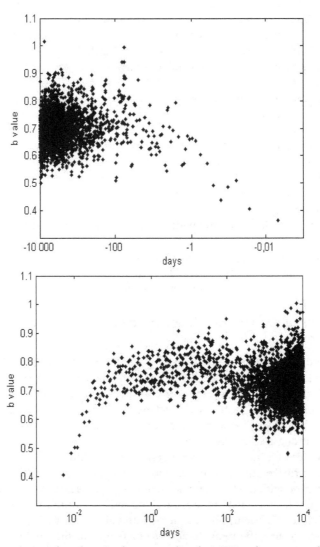

Fig. 3. The change of mean b-values in the generalized vicinity of strong earthquake. The values obtained for 50 events groups are given by dots. Panels as in Fig. 1.

aggregation in 0.1-bins is small, it influences the b-values weakly and uniformly; therefore it was not taken into account. The maximum likelihood method (3) gives a suitable b-value estimation for a number of events exceeding 50. Having this in mind the groups consisting of 50 subsequent events were used in b-value determination. The data points in Fig. 3 reflect the b-values obtained for such groups with step 25 events; thus the data points are independent of those next to the adjacent ones.

As it can be seen in Fig. 3, there is an evident tendency of decrease in b-values in the time vicinity of the generalized main shock; and this decrease increases strongly with approaching the moment of the main shock. In the foreshock sequence the noticeable decrease begins about one hundred days before the main shock. In the aftershock sequence the sharp increase in b-values takes place during the first several days after the main shock. A slow increase in b-values takes place in the following 100 days. It is necessary to notice that the b-values appear to be increased in comparison with the background value in the time interval 10-100 days after the main shock occurrence. These features agree with a tendency of lowermost b-values in the very beginning of the aftershock sequences and with an increase of b-value in the further evolution of the aftershock sequences (Rodkin, 2008; Smirnov & Ponomarev, 2004). The similar tendency was found in the examination of acoustic emission data (Smirnov & Ponomarev, 2004). New findings consist in the stronger decrease than it was found before and in rather symmetrical character of this decrease for fore- and aftershock sequences. Note that the amplitude of the b-value decrease appears to be proportional to the logarithm of time remaining from the moment of the main shock. Such type of behavior is typical of critical processes.

Note however that the well known and widely used below effect of "seismic quiescence" was not found in the generalized vicinity of strong earthquake. It can be connected with anisotropic character of this type of precursor anomaly in relation to a strong earthquake epicenter that is mentioned in (Zavyalov, 2006). In this case this effect can be eliminated by summarizing data from vicinities of a large number of differently oriented strong earthquakes.

4. Experience in earthquake prediction at the Sakhalin Island and surrounding areas

Region under study includes the Sakhalin Island and the Kuril Islands arc. In a few cases the area of the Japan Islands was also taken into account. This territory belongs to the transitive zone between the Pacific and the Eurasian continent and includes the active island arc characterized by one of the highest levels of seismicity on the Earth. Because of variability in quality of available catalogs the methodology of prognosis is more or less different in every particular case of strong earthquake prognosis, which is described below.

To avoid misunderstanding and controversial interpretations, we follow below the definition of the term "earthquake prediction," which was formulated by the Panel on Earthquake Prediction with the US National Academy of Sciences (Allen et al., 1976):

"An earthquake prediction must specify the expected magnitude range, the geographical area within which it will occur, and the time interval within which it will happen with sufficient precision so that the ultimate success or failure of the prediction can readily be judged. Only by careful recording and analysis of failures as well as successes can the eventual success of the total effort be evaluated and future directions charted. Moreover, scientists should also assign a confidence level to each prediction." (Predicting earthquakes ..., 1976).

In the case when the demands of this definition are not fulfilled the term "earthquake forecasting" is used.

We use such definition instead of another one when the term "prediction" means a deterministic prognosis as it was formulated in (Operational Earthquake Forecasting: State of Knowledge and Guidelines for Utilization. International Commission on Earthquake Forecasting for Civil Protection, http://www.protezionecivile.gov.it/cms/attach/ ex_sum_finale_eng1.pdf). We suggest that deterministic prognosis is impossible now and hardly will be possible even in future, thus such use of the term "prediction" seems to be inefficient.

4.1 Case 1 - Diagnostics of a dangerous period before the 1994 Mw 8.3 Shikotan earthquake (the South Kuril region)

4.1.1 Seismic region and data

The region under study in case 1 includes the Kuril Islands zone and the area to the east of Hokkaido Island. In 1992 we have prepared in computing form the earthquake M ≥ 4.0 catalog of the Kuril-Okhotsk region for the period 1962-1990. It was formed on the basis of yearly publications (The earthquakes in USSR..., 1964-1991). During the next years the catalog was updated by the Operative catalog data of the Sakhalin Branch of Geophysical Survey of the RAS. We used this regional catalog for testing the M8 algorithm (Keilis-Borok & Kossobokov, 1986, 1990) which provided a suitable procedure for prediction of large earthquakes.

4.1.2 Methodology

The intermediate-term earthquake prediction technique, named M8 algorithm, is based on an assumption that a number of functions, defined for a particular earthquake sequence, become extremely large in values, within several months prior to a major shock. The functions used are following (Keilis-Borok & Kossobokov, 1986, 1990):

N – cumulative number of main shocks (aftershocks are excluded according to (Keilis-Borok et al., 1980)) describes an increase in seismic activity;

L – describes deviation of N from the long-term trend value;

Z – describes a linear concentration of earthquake sources;

B – describes the bursts of aftershocks.

All functions, except the last one, were calculated twice: for a standard variant of small statistics (10 events or less per year) and for a standard variant of large statistics (20 events or more per year); where the numbers of events change by choice of threshold of magnitude taken into account. Two statistics are used for increasing robustness of results of prognosis. Values of these seven functions were used for adjusting the M8 algorithm, and then for diagnostics of Time of Increased Probability (TIP) for large earthquake (M ≥ 7.5) occurrence within the circular areas with a fixed radius.

Besides the method described above, we used a visualization technique to display space-time distribution of seismicity to detect seismic gaps of the second kind. A gap of the second kind (seismic quiescence) refers here to a portion of a seismic area of low seismic activity with no observed earthquakes with M≥6.0 for a period of several years. This approach follows the concept of K. Mogi (1985).

4.1.3 Results of analysis and precursors phenomena

Seismicity of circular areas with a radius of 427 km with the centers located in the points: 44^0 N, 149^0 E; and 48^0 N, 155^0 E has been examined. These two circles overlap all the territory of

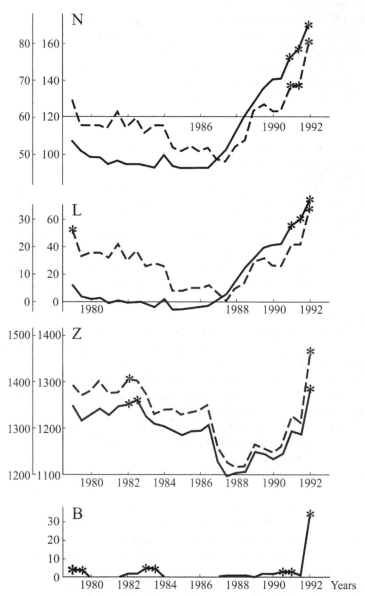

Fig. 4. The behavior in time of seven functions of the M8 algorithm for the Southern Kuril region during the diagnostics period, 1979-1992. The solid lines show the values, calculated for large statistics (20 events or more per year), and the dash lines show the values for small statistics (10 events or less per year). Star symbols mark anomalous values.

the Southern and Northern Kuril Islands. The catalog data have been processed by the M8 algorithm for the time period from 1962 to July 1992, with the functions being calculated for every six months.

Some results of the processing are presented in Fig. 4, which demonstrates the behavior of all seven functions in the first circular area (Southern Kuril Islands) during the diagnostics period (1979-1992). All the functions have become extremely anomalous, large in values to the July of 1992, which means, that the M8 algorithm diagnoses the TIP for a large earthquake occurrence during next 5 years (1993-1997). The alarm should be kept if anomalous values of almost all the functions are kept in the next six months.

The similar results have been derived by the authors of the M8 algorithm on the base of processing of the NEIC/USGS catalog data (Kossobokov et al., 1994, 1996). All needed parameters of the prognosis of the future strong earthquake were indicated, and thus we suggest that in case 1 the term "prediction" is suitable.

For the second circular area (Northern Kuril Islands) anomalous value has been obtained for the B function (bursts of aftershocks) only, and it means, that M8 algorithm diagnoses no TIP for a large earthquake occurrence in this area within the next 5 years.

The above mentioned suggested a high probability of occurrence of large earthquake within the Southern Kuril zone in the nearest years. This suggestion was found to be in an agreement with the space-time distribution of earthquakes with M ≥ 6.0 within the Kuril seismic zone since 1987 (Fig. 5). A large seismic gap of the second kind can be seen within a big area from the southern part of Urup Island to the northern end of Hokkaido Island.

4.1.4 Realization of prediction

The prediction described above was submitted in July of 1992 to the Russian Academy of Sciences and the Ministry of Emergency Situations (REC RAS/EmerCom). It was written in the conclusion that "the Southern Kuril region and the area to the east of Hokkaido Island will remain in a state of high probability of a large (M=7.5-8.5) earthquake occurrence during 5 years, which started since the middle of 1992" (Kossobokov et al., 1994, 1996).

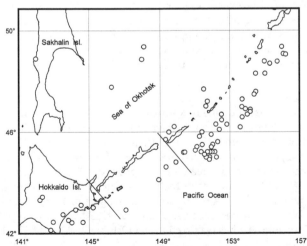

Fig. 5. The distribution of epicenters of earthquakes with M ≥ 6.0 in the Kuril-Hokkaido area for the period from March 1987 to July 1992. The area limited by solid lines is the area seismic gap of the second kind.

A large Mw 8.3 shallow-focus (h ~ 40 km) earthquake has occurred on 04 October 1994 at 13:22 GMT to the east of Shikotan Island (Russia) (Fig. 6). Thus, the intermediate-term prediction of July 1992 was confirmed.

Fig. 6. Epicenters of the 1994, Mw 8.3 Shikotan earthquake and first-day aftershocks of magnitude M ≥ 5.0.

Just before the Mw 8.3 Shikotan earthquake the seismic stations in Kurilsk and at Shikotan Island were closed because of the economic crisis. In this situation we have no data to attempt to perform a short-term prognosis. But using a posteriori data from USGS/NEIC a short-term prognosis of this event was done. We used the method of self-developing processes, which was suggested by Malyshev (Malyshev, 1991; Malyshev et al., 1992). It is described below (case 4) where it was applied in a real time. By the use of this method the one and a half year foreshock sequence of events was analyzed and the date of the strong earthquake occurrence was a posteriori estimated with a few days delay (Fig. 7).

4.1.5 Case 1 summary

Some characteristics of the earthquake flux for the period from 1962 to July 1992 in the Kuril seismic zone have been investigated on the basis of two methods: (1) the intermediate-term earthquake prediction algorithm M8; (2) a visualization of space-time distribution of seismicity. The M8 algorithm diagnosed the Time of Increased Probability for a large earthquake occurrence in the circular area with the radius of 427 km at the point (44⁰ N, 149⁰ E) during the period 1993-1997. By means of the second method the seismic gap of the second kind was detected within a big area from the southern part of Urup Island to the northern end of Hokkaido Island. The quiescence began in March 1987. A catastrophic shallow-focus (h ~ 40 km) Mw 8.3 earthquake has occurred on 04 October 1994 at 13:22 GMT to the east of Shikotan Island (Russia).

A posteriori short-term prognosis by the method of self-developing processes data was performed using USGS/NEIC data. The date of the strong earthquake occurrence was a posteriori estimated with a few days delay (Fig. 7).

Current State of Art in Earthquake Prediction, Typical Precursors and Experience in Earthquake Forecasting at Sakhalin
Island and Surrounding Areas

55

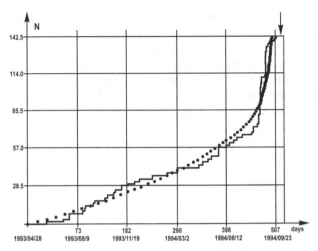

Fig. 7. The cumulative number of earthquakes observed by USGS/NEIC in the Southern
Kuril Islands area as a function of time through April 1993–November 1994.

The stair-case curve is empirical data, and the smooth curve simulates the data according to
the method of self-developing processes. The vertical line is the asymptote corresponding to
the prognostic event date three days after the earthquake occurrence (the arrow).

4.2 Case 2 - Partly retrospective forecasting of the May 27, 1995 Mw 7.1 Neftegorsk earthquake, North-Eastern part of Sakhalin Island, Russia
4.2.1 Seismic region and data
Seismicity of the northern part of the Sakhalin region (north of latitude 50^0 N) was the object
of this investigation. It had a moderate level of seismicity in comparison with seismic
activity of the Kuril Islands. Destructive earthquakes like the 1995 Mw 7.1 Neftegorsk
earthquake are rare events here. Paleoseismological reconstruction within the Upper Piltun
fault, which was reactivated during the Neftegorsk earthquake, showed that recurrence of
such earthquake is about one event per several hundred years (Shimamoto et al., 1996).
Seismicity patterns were analyzed on the basis of the regional catalog of shallow-focus M ≥
3.0 earthquakes, issued by ESSN (The earthquakes in USSR…, 1964-1991).

4.2.2 Methodology
In this case same methods as in the case of the Shikotan earthquake were used. A magnitude
for identification of a seismic quiescence area for the Sakhalin region was taken M = 3.0.

4.2.3 Results of analysis and precursors phenomena
The second kind seismic gap area taking place along the eastern coast of the Northern
Sakhalin has indicated the approximate location of a possible future large earthquake (Fig.
8) (Kim, 1989). The gap of the second kind was recognized in 1989, i.e. 6 years before the
Neftegorsk earthquake, it was outlined in the area of 200 by 60 km including the shelf and
coastal areas from the southern part of the Shmidt Peninsula to the Gulf of Chaivo. There
were no earthquakes with M ≥ 3 in this area since 1984.

We have confirmed the existence of the quiescence zone in (Saprygin et al., 1993). In this paper we advised to reinforce the Northern Sakhalin network of seismic monitoring. However, in this very time because of the economic problems in this country four seismic stations from six, which controlled this region, were closed down. A large number of objects of industrial and civilian purposes were built with the reference seismicity of 6-7 of the MSK-64 intensity scale. Thus, there was a deficit of seismic-resistant buildings and structures. That became evident when the May 27, 1995 Neftegorsk earthquake has occurred. The observed ground shaking intensity was 8-9 (MSK-64 scale) in Neftegorsk, and 1841 inhabitants were killed (A memory ..., 2000; Streltsov, 2005).

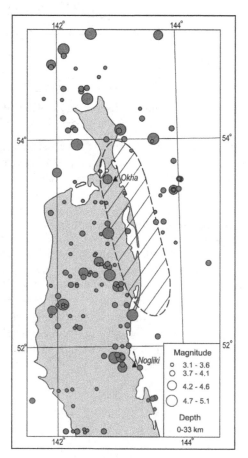

Fig. 8. A seismic quiescence zone (hatched area) in the northern region of Sakhalin Island recognized on the basis of absence the magnitude M ≥ 3.0 events since July 1984 (Kim, 1989). The map shows a state of seismicity in April 1988.

We have investigated the intermediate-term precursors of the Neftegorsk earthquake by means of the M8 algorithm (Keilis-Borok & Kossobokov, 1986, 1990; Tikhonov, 2000). It was applied for the retrospective diagnostics of TIP for this earthquake. We used the declustered

Current State of Art in Earthquake Prediction, Typical Precursors and Experience in Earthquake Forecasting at Sakhalin Island and Surrounding Areas

57

regional catalog (Keilis-Borok et al., 1980). The M8 algorithm was adjusted to the earthquake catalog for the period 1964 – 1978. A dangerous period was found after 1979 (Fig. 9).

In this case data processing was performed under a strong shortage of data (only 4-5 events per year). Dr. Kossobokov analyzed two cases in the similar poor data conditions for the use of the M8 algorithm - deep earthquakes of the Vrancha region (Kossobokov, 1986) and seismicity in Greece (Latoussakis & Kossobokov, 1990). In the first case small and "large" statistics was equal to 2 and 4 events per year but in the second case it was equal to 5 and 10 events per year. However, even under such unfavorable conditions the M8 algorithm has demonstrated an ability to recognize the danger.

In our case only one dangerous period was revealed a posteriori since 1991 when six functions became anomalous (B function was undefined because of poor statistics of small earthquakes). The alarm period was interrupted by the May 27, 1995 Mw 7.1 Neftegorsk earthquake (Fig. 10).

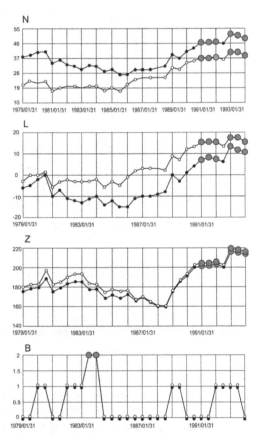

Fig. 9. The behavior in time of seven functions of the algorithm M8 for the northern region of Sakhalin Island during the diagnostics period (1979-1993) before the May 27, 1995 Mw 7.1 Neftegorsk earthquake (Tikhonov, 2000). The white circles show the values, calculated for "large" statistics (5 events or more per year), and the black circles show the values for small statistics (4 events or less per year). Large black circles mark anomalous values.

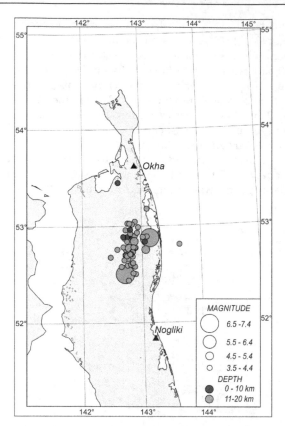

Fig. 10. Epicenters of the May 27, 1995 Mw 7.1 Neftegorsk earthquake and first-day aftershocks of magnitude M ≥ 3.5.

4.3 Case 3 - Incomplete forecasting of the Tokachi-oki Mw 8.3 earthquake (Hokkaido Island, Japan)

The M8 algorithm has failed in prognosis of the Tokachi-oki Mw 8.3 earthquake (http://mitp.ru/predictions/html, this site is of access for experts only since 2000 year). Despite of the failure of the M8 algorithm, the described below ZMAP-technique performed by one of the authors was successful (Tikhonov, 2003; 2005).

4.3.1 Seismic region and data

In this case the territory of the Japanese Islands including the adjacent shelf areas (Fig. 11) was examined. The Japan Meteorological Agency earthquake catalog from January 1974 until July 2002 was used. Earthquakes with M ≥ 3.8, H ≤ 100 km were found to be completely recorded, and these events were taken into account (Fig. 11). This data set is quite homogeneous throughout the whole region of Japan. It permits to apply the ZMAP-technique (Wiemer & Wyss, 1994) for examination. This method could not be applied in the cases 1 and 2 because of a shortage of data and difference in data availability for the Northern and Southern areas of the Sakhalin Island.

4.3.2 Methodology

The ZMAP method (Wiemer & Wyss, 1994) was developed to reveal a change in rate of seismicity as a function of space and time. The authors have used a rectangular grid with a spacing of 2 km for the total studied area about 100 by 100 km. For each grid point N_i nearest epicenters are selected and the maximum distance of an earthquake from the i-th grid point $r(N_i)$ was calculated. Thus the defined $r(N_i)$ is a function of space proportional to the local density of earthquakes. The significance of change in seismicity rate for each grid point is evaluated using the standard z test (Habermann, 1981, 1982)

$$z(t) = (R_{all} - R_{wl}) / (\sigma^2_{all} / n_{all} + \sigma^2_{wl} / n_{wl})^{1/2}, \tag{4}$$

where R_{all} и R_{wl} are the mean rates of seismic process in all observation period (from t_o to t_e) and in sliding window wl, respectively. n indicates the number of samples, σ is the standard deviation.

To visualize the changes in the rate of seismicity the authors plotted z(t) values on a map. Moment t moves through the whole period of the catalog from t_o to t_e. To identify the strongest rate changes between two intervals (from t_o to t and from t to t_e) they have used the $AS(t)$ function (*Habermann, 1983, 1987, 1991*). This function gives the most probable moment of seismic quiescence occurrence:

$$AS(t) = (R_1 - R_2) / (\sigma^2_1 / n_1 + \sigma^2_2 / n_2)^{1/2}, \tag{5}$$

where R_1, R_2 are the mean rates of seismic process in two periods (from t_o to t and from t to t_e), n_1 and n_2 are the numbers of samples in these periods, σ_1, σ_2 are the standard deviations in these periods.

In the process of application of the ZMAP-technique the following tasks were executed for detection of seismic quiescence periods in the Japan region (Tikhonov, 2003, 2005):

- A modification of the ZMAP-method for application to a large territory in a real time scale has been executed. After the modification the task was implemented using the standard deviate z test (Habermann, 1981, 1982) in two steps: (1) Detection of seismic quiescence in a studied region using a coarse rectangular grid with a moderate number of nodes (with a spacing of 0.25^0); (2) Covering the cells where seismic quiescence was detected by a detailed grid (with a spacing of 0.1^0) and calculation of a configuration of anomalous area with a given value of seismicity rate decrease.
- An adjusting of the modified ZMAP-technique to the JMA earthquake catalog for the detection of possible seismic quiescence periods before the strong shallow earthquakes with $M \geq 6.8$, $H \leq 100$ km.
- An investigation of the precursor seismic quiescence since July 2001 within the studied area.

4.3.3 Results of analysis and precursors phenomena

In order to effectuate the first step of methodology we divided the studied territory into grids spacing 0.25^0 in latitude and longitude (Fig. 11). An adjustment of a modified method was performed to the declustered earthquake catalog for the period 1975 – 1988. The values of z(t) function were identified as anomalous if they exceeded a proper threshold calculated for the adjusting time span. Thresholds U_i for detection of quiescence in separate nodes was defined in the following way:

$$U_i (coef) = \mu_i + coef \ \sigma_i, \tag{6}$$

where μ_i, σ_i are the average and the root mean square values of function $z(t)$ in node i for the learning time, respectively; *coef* is empirical constant. The value of $z(t)$ were identified as anomalous if $z(t) \geq U_i$ *(coef)*. The constant *coef* was taken equal to 4. Thresholds for nodes were selected to minimize a probability of omission of a real seismic quiescence prior the strong earthquakes with $M \geq 6.8$.

Detection of the areas with anomalous values of $z(t)$ function has been fulfilled for the declustered catalog data since 1989. Thus, there were facilities for detection of seismic quiescence periods occurring prior a series of large seismic events, which occurred in 1992 – 2002. As a result, we obtained a set of maps of the Japan region showing the location of such areas at different moments of time. Dynamics of the appearance and evolution of the anomalous areas was compared visually with dynamics of the occurrence of the strong earthquakes ($M \geq 6.8$, $H \leq 100$ km). It was established that correlation between the most outstanding anomalies and the strong earthquakes was suitable in space and time. In general the maximum size of anomaly is observed about 0.5 – 1.5 yr before the corresponding strong shock. The results of processing of the catalog since 1989 were the following: in 7 cases the occurrence of the strong seismic events was forestalled by seismic quiescence near its epicenters. In general the epicenter is located near the border of the corresponding anomalous zone. In two cases there was no quiescence before the strong earthquakes, and in two cases anomalous areas were observed before swarms of moderate size earthquakes ($M = 6.2 – 6.6$).

Obviously, the recent seismic quiescence zone revealed in the northern part of Japan had attracted an interest (Fig. 12). The term "recent" dates here back to the time of investigation (the middle of 2002). As a result of the second step of the procedure (with a detailed spacing of 0.1^0 grid size) the most outstanding recent anomaly of 75 km size was located near the Cape Erimo (Hokkaido Isl.) (Fig. 13). It was characterized by a seismicity rate decrease of 75% starting from January 1998. Inside this anomalous area there was a circle with R=25 km with no earthquake occurrence with $M \geq 3.8$, $H \leq 100$.

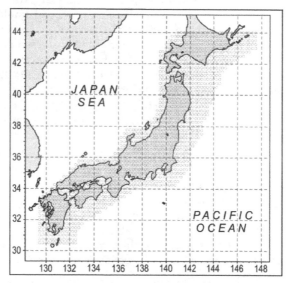

Fig. 11. Map displaying the grid with spacing of 0.25^0 used for detection of seismic quiescence. This grid contains 1354 nodes.

Current State of Art in Earthquake Prediction, Typical Precursors and Experience in Earthquake Forecasting at Sakhalin
Island and Surrounding Areas

61

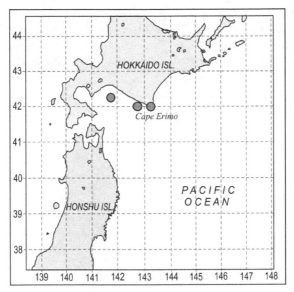

Fig. 12. The location of the seismic quiescence in the northern part of Japan as on June 1, 2002. Gray circles denote the anomalous nodes. The anomaly of seismic quiescence started in January 1998.

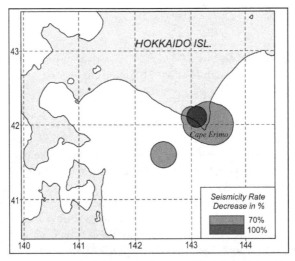

Fig. 13. Map of the seismicity rate decrease calculated for the grid with spacing of 0.1⁰ at the middle of 2002. The seismic quiescence anomaly within the Cape Erimo started in January 1998.

4.3.4 The realization of forecasting

The Tokach-oki earthquake forecasting was presented during the XXIII General Assembly of IUGG, which was held in Sapporo (June 30 – July 11, 2003) (Tikhonov, 2003). Besides, these results were published in (Tikhonov, 2005). The manuscript of the paper was received by the Journal of Volcanology and Seismology on 6 August 2003, i.e. before the occurrence of the Tokachi-oki earthquake. The Tokachi-oki earthquake Mw 8.3 occurred on September 26, at 4 h 50 min JST time near the southern coast of Hokkaido close to the seismic quiescence zone (Fig. 14).

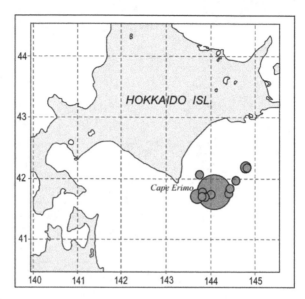

Fig. 14. Map of the 26 September, Mw 8.3 Tokachi-oki earthquake epicenter (large circle) and its first-day aftershocks of magnitude M≥5.0 (small circles).

4.3.5 Case 3 summary
The modified ZMAP-method has been applied to detect precursory seismic quiescence zones in the Japan region. The anomalies revealed for the period 1989 – 2000 correlate in space and time with the strong event occurrences. The greatest size of anomalous area took place typically about 0.5 – 1.5 yr before the corresponding strong shock.

The anomalous decrease of shallow seismicity (M ≥ 3.8) was detected in the southern part of Hokkaido islands at the middle of 2002. In result of the second stage of the procedure the anomaly of 75 km size was determined. It was characterized by a seismicity rate decrease of 75% from January 1998. Moreover, inside this zone there was a circle of 25 km radius with 100% decrease of the rate. The Tokachi-oki earthquake Mw 8.3 has occurred on September 26, at 4 h 50 min JST time close to the seismic quiescence zone (Fig. 14).

4.4 Case 4 - A successful prediction of the 2 August, 2007 Nevelsk earthquake (Mw 6.2) in Southern Sakhalin Island
4.4.1 Seismic region and data
The object of this investigation was the southern part of the Sakhalin Island (south of latitude 48°N). The basic feature of the Earth's crust in this region is characterized by close

Current State of Art in Earthquake Prediction, Typical Precursors and Experience in Earthquake Forecasting at Sakhalin
Island and Surrounding Areas

63

arrangement of three major fault systems marked by recent seismic activity. These are the Rebun–Moneron, the Western, and the Central Sakhalin fault zones (Fig. 15).

We used two data sets in this study: (1) the catalog of shallow earthquakes for 1992–2002 from the IRIS-2 system, installed at the "Yuzhno-Sakhalinsk" seismic station in 1992 (Kraeva, 2003); (2) the catalog of network of digital "Datamark" and "DAT" autonomous seismic stations, operating since 2001. The first catalog provides a record of all M > 2.6 seismic events within epicenter distances up to 70 km from the station. The second catalog is more detailed and provides analysis of seismicity patterns in the whole southern part of Sakhalin Island.

4.4.2 Methodology

This prediction was based on the detection of seismic gaps of the first and second kind. Let us describe these terms for the examined situation of the moderate seismic activity of the Sakhalin Island in detail. A gap of the first kind refers to a portion of a seismic area that has been in a state of relative rest for a long time (100 years and more), i.e., there have been no earthquakes with magnitude M ≥ 6.0 during this period. A gap of the second kind (seismic quiescence) refers to a portion of a seismic area of low seismic activity with no earthquakes with M ≥ 3.0 observed for a period of several years. Note that in the case 1 the second kind gap was examined for the magnitude threshold M = 6.0 because of the higher seismic level of the Kuril Islands.

We have used also the method of self-developing processes suggested by Malyshev (Malyshev, 1991; Malyshev et al., 1992). It was found that behavior of empirical earthquake sequences before and after large seismic events is satisfactory described by solutions of a nonlinear differential equation of the second order:

$$\frac{d^2x}{dt^2} = k \left| \left(\frac{dx}{dt} \right)^2 - V_0^2 \right|^\gamma , \tag{7}$$

where x is a parameter of process (for example, a cumulative sum of a number of shocks – N parameter), $V_0 = (dx/dt)_0$ is a rate of seismic process in stationary state, k and γ are empirical constants. Particular solution of the equation in case of $2\gamma > 1$ has a vertical asymptote. The time position of this asymptote is shown to be close to the origin time of the ongoing strong earthquake.

4.4.3 Precursors phenomena and characteristics of prediction

Apparently, each of three above mentioned fault zones has the potential to originate major earthquakes Ms 7.0–7.5. However, evidence is currently limited to the Rebun–Moneron (the 1971 Moneron earthquake, Ms 7.5) and the Central Sakhalin (paleoseismological data) fault zones. The Western Sakhalin fault zone showed no magnitude M > 5.0 events in its southern part during the whole history of instrumental observations up to 2006 (Fig. 15). However in its northern part (latitude>48°N) it has originated large earthquakes in 1907 (Alexandrovsk–Sakhalinsk, Ms 6.5), 1924 (Lesogorsk–Uglegorsk, Ms 6.9), and 2000 (Uglegorsk, Ms 7.2) (Fig. 16). Besides these three major fault zones in the studied area there are a number of small fault zones of lower seismic potential.

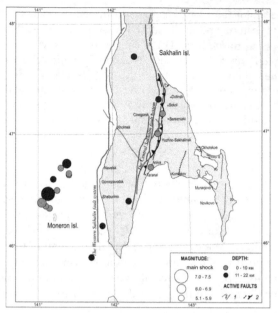

Fig. 15. Map of crust earthquake M > 5.0 epicenters of southern Sakhalin, 1906–2005, and the main fault zones.
Notes: The active faults are plotted according to M.I. Streltsov of IMG&G FEB RAS, Yuzhno-Sakhalinsk (1) and A.I. Kozhurin, of GIN AS, Moscow (2).

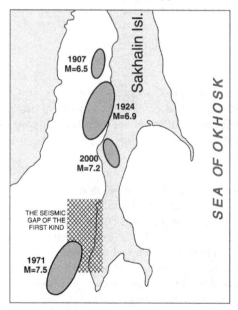

Fig. 16. Sources of large earthquakes at the western coast of Sakhalin Island (grey ovals) and the approximate location of the seismic gap of the first kind (hatched rectangle).

Current State of Art in Earthquake Prediction, Typical Precursors and Experience in Earthquake Forecasting at Sakhalin
Island and Surrounding Areas

65

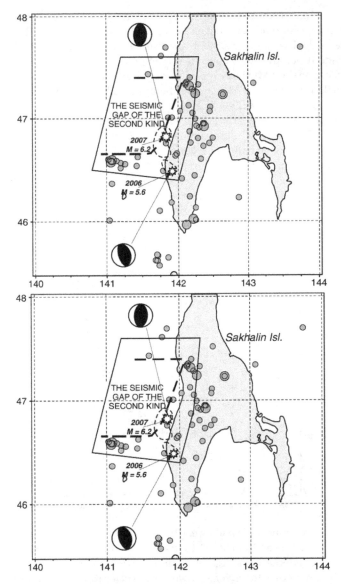

Fig. 17. Map of crust earthquake M≥3.0 epicenters, 1993–2005, recorded by the "Datamark"
network of stations and the "IRIS-2" system, installed at the "Yuzhno–Sakhalinsk"
seismic station.
Note: the area of the seismic gap of the second kind is outlined with the bold dash line,
while the source zones of the 17 August 2006, Mw 5.6 Gornozavodsk and the 2 August 2007,
Mw 6.2 Nevelsk earthquakes, with a thin dash line; asterisks indicate the epicenters of main
shocks; focal mechanisms are given based on data from [http://www.globalcmt.org]. The
area limited by the polygon is the geographical area within which a large earthquake
M=6.6±0.6 may occur.

An earlier publication (Tikhonov, 1997) recognized an incipient of the second kind gap (seismic quiescence) within one of the gaps of the first kind, situated on the western coast of southern Sakhalin. In December 2005 our analysis of the southern Sakhalin network data permitted to made it possible to: (1) outline rather precisely the area of a seismic gap of the second kind, where shallow earthquakes with magnitudeM ≥ 3.0 did not occur from at least the middle of 2003 (Fig. 17); and (2) observe the appreciable revival of seismic activity that eventually encircled this area by 2003 (Fig. 18).

Furthermore, the rise of activity around the seismic quiescence zone, and the area south of it, has accelerated (Fig. 18) with culminations linked to the 30 May 2004 Kostroma, Ms=4.8 earthquake in the Western Sakhalin fault zone and the 18 December 2004 Moneron, Ms=4.7 earthquake in the epicenter area of the major 1971 Moneron earthquake. This happened while the seismic sequence of the abovementioned 2001 Takoye earthquake swarm in the Central Sakhalin fault zone was still ongoing.

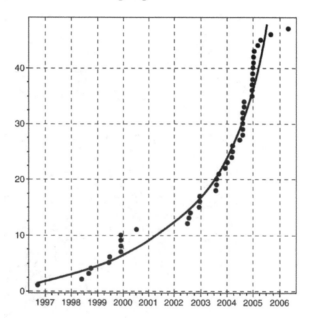

Fig. 18. The cumulative number of shallow earthquakes (depth above 30 km) of magnitude 3 or more inside 45.5–46.75°N and 140.8–142.2°E (i.e., next to the southern area of the identified seismic gap of the second kind) as a function of time in September 1996–May 2006.
Note: the smooth line models the inverse power law acceleration of the empirical data according to the method suggested by Malyshev et al. (1992). The line has the asymptote at 26 August 2007.

At the time, Dr. Tikhonov, in collaboration with Ch.U. Kim, A.I. Ivashchenko and L.N. Poplavskaya (Institute of Marine Geology and Geophysics, Yuzhno–Sakhalinsk) issued the long-term prediction of major earthquake near the western coast of southern Sakhalin (Tikhonov, 2006). This strong earthquake prediction summarized in Table 1 was made by taking into account 1) the seismic gap of the first kind in the Western Sakhalin system of faults, where large earthquakes were absent for at least 100 years (Fig. 16); 2) the seismic gap

Current State of Art in Earthquake Prediction, Typical Precursors and Experience in Earthquake Forecasting at Sakhalin
Island and Surrounding Areas

67

of the second kind in the area of 90 by 60 km where seismic quiescence was confirmed by accurate data from "Datamark" digital network (Fig. 17); and 3) the accelerated sequence of earthquakes in the area adjacent to it (Fig. 18). The prediction was the following:

- The location of the incipient hypocenter is most likely to occur at shallow depths within 0–30 km inside the polygon (Fig. 17): (47.6° N; 141.2° E); (46.5° N; 140.8° E); (46.4° N; 142.0° E); (46.9° N; 142.2° E); (47.6° N; 142.3° E).
- The magnitude Ms of the incipient event was estimated in two different ways. The first is the formula by K. Tanaka (1980) lg R=0.33 M−0.07, which relates the linear size of the gap of the second kind, R, to the magnitude of expected earthquake, M. It was used first to determine the expected magnitude M=6.1.
- The second estimate was obtained from the two empirical relations: (1) lg L=(0.5±0.01) M−(1.77±0.07) (Tarakanov, 1995); and (2) L ≈ 1/3 R, where L is the linear size of the aftershock zone (Shebalin, 1961). Substituting an expression (2) in the formula (1), we obtain:

$$lg\ R=(0.5\pm0.01)\ M-(1.77\pm0.07)+lg\ 3. \qquad (8)$$

This gives M=6.6.

- Of the two estimates, the second appears to be preferable because it takes into account the worst earthquake scenario as well as some uncertainty in estimates. Therefore, Ms=6.6±0.6 was selected as a final magnitude estimate of the expected large earthquake.
- The duration of alarm was determined to be about 7.5 years. This was based on an average time-span of approximately 10 years, observed for seismic quiescence zones that occurred before large earthquakes off the western coast of Japan and Sakhalin, while accounting for no less than 2.5 years of a given quiescence zone's initiation.
- The likelihood of an earthquake occurrence was estimated at 75%, based on the recurrence rate of the large (M ≥ 6.5) earthquakes in the south of Sakhalin (Oskorbin & Bobkov, 1997) and the lifespan of a given quiescence zone.
- The expected intensity of ground shaking (in the MSK-64 scale) was calculated for the three epicenter locations inside the seismic gap of the second kind and the magnitude close to the maximal expected. Fig. 19 displays the results obtained with the epicenter in the middle of the quiescence zone.

The beginning and end of alarm	Magnitude and depth of earthquake	Position of earthquake epicenter	Probability of earthquake occurrence	Maximal macroseismic effect (MSK-64 scale)
January, 2006–July, 2013	$M_S = 6.0 - 7.2$ h = 0 – 30 km	See the text and Fig. 17	≥75%	9.0 (in epicenter) 8.0 (at the coast)

Table 1. Characteristics of anticipated earthquake (Tikhonov, 2006, page 179).

The prediction described was submitted in January 2006 to the Russian Expert Council for Earthquake Prediction, Seismic Hazard and Risk of the Russian Academy of Sciences and the Ministry of Emergency Situations (REC RAS/EmerCom). As a result of the discussion at the REC Meeting, the prediction was approved as being scientifically motivated. It was then reported to EmerCom headquarters, which had run urgent command-staff exercises in August 2006, referred to as "Mitigating the consequences of destructive earthquake and tsunami in Sakhalin–Kuril region." The scientific motivations of the prediction have been published (Tikhonov, 2006).

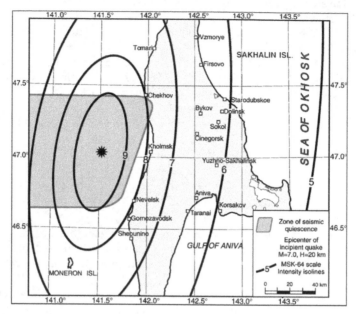

Fig. 19. An expected ground shaking intensity (MSK-64 scale) for the model occurrence of an earthquake of Ms 7.0 at a depth of 20 km in the central part of the seismic quiescence zone (computations by L.N. Poplavskaya of IMG&G FEB RAS made in December 2005).

4.4.4 The realization of the prediction

On 17 August 2006, a magnitude Mw 5.6 earthquake hit the Gornozavodsk settlement in the area of prediction (Levin et al., 2007). Upon analyzing the main shock and its aftershocks, a conclusion of the precursory character of this sequence was drawn. Specifically, it was concluded that the preparation of a large earthquake in the seismic quiescence zone had switched from a long-term to a short-term phase.

This was briefly formulated in the cover letter of an interim report to REC RAS/EmerCom, as follows: "In December 2005, the seismology team of IMG&G FEB RAS issued a long-term prediction of strong earthquake on southwest shelf of Sakhalin Island. Recent M 5.6 earthquake, which happened in this region on 17 (18) August 2006, has partially proved the prediction to be well-founded. Detailed analysis of post-earthquake seismicity allows to conclude that the development of predicted earthquake is in the short-term stage now…".

On August 2, at 13 h 37 min Sakhalin time (2 h 37 min GMT), in the Tatar Strait close to the city of Nevelsk (Sakhalin, Russia), an earthquake of magnitude Mw 6.2 occurred (Fig. 20). Two lives were lost and more than ten persons were wounded. The earthquake caused severe destruction. About six thousand of Nevelsk's fifteen thousand inhabitants became homeless. The earthquake was felt everywhere in the southern portion of Sakhalin Island. The observed groundshaking intensity (MSK-64 scale) was VII–VIII in Nevelsk, VI–VII in Gornozavodsk, V–VI in Holmsk and III–IV in Yuzhno-Sakhalinsk.

Current State of Art in Earthquake Prediction, Typical Precursors and Experience in Earthquake Forecasting at Sakhalin Island and Surrounding Areas

69

Inspections into the consequences of this disaster have shown that the city needs to be rebuilt practically anew. The losses totaled more than six billion rubles (i.e., $240 million). The focal mechanism of the main shock, based on data from (http://www.globalcmt.org), suggests that the source region was under the sub-latitudinal and near-horizontal compression that resulted in the reverse-slip (Fig. 17). IMG&G and employees of the Sakhalin Branch of Geophysical Survey of the RAS carried out a general inspection of the region affected by the earthquake. Other organizations provided the aerial mapping and echo sounding of the sea-bottom. The seismic event appeared to be related to the West-Sakhalin system of deep crustal faults located along the western coast of the island. As a result of the general inspection, a number of unique observations for earthquakes of such size have been established. One of the most remarkable geodynamic phenomena associated with the 2007 Nevelsk earthquake is the uplift of the coastal terrace, formed by the Middle Miocene sedimentary rocks (Nevelsk suite), with an amplitude of 1.0–1.5 m (Fig. 21).

The 2 August 2007, Mw 6.2 Nevelsk earthquake occurred in the southern part of the seismic gap of the second kind (Fig. 17). Its parameters fall within the limits of the long-term prediction of a large earthquake expected in the southwest of Sakhalin Island, as it was listed in the Table 1.

Thus, the long-term prediction of December 2005 was confirmed. Note also that the decision that the 17 August 2006 Gornozavodsk earthquake was a foreshock of a future large event was declared just after this event (23 August 2006). More details concerning case histories of prediction of the 2006 Gornozavodsk and the 2007 Nevelsk earthquakes can be found in (Levin et al., 2007; Tikhonov & Kim, 2010).

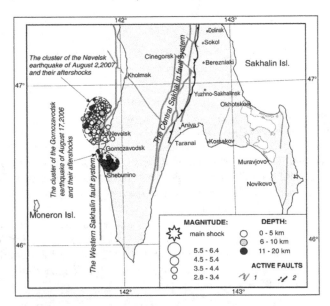

Fig. 20. Map of the 17 August 2006, Mw 5.6 Gornozavodsk and 2 August 2007, Mw 6.2 Nevelsk earthquake epicenters (asterisks) and their first-day aftershocks of magnitude M ≥ 2.8. Notes: the clusters of epicenters are outlined with a dash line. The active faults are plotted according to M.I. Streltsov of IMG&G FEB RAS, Yuzhno-Sakhalinsk (1) and A.I. Kozhurin, of GIN AS, Moscow (2).

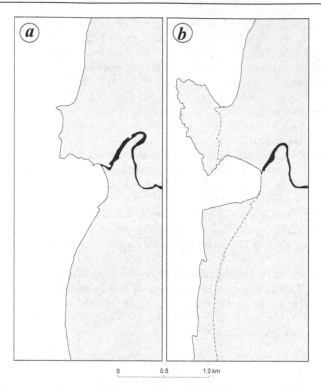

0   0.5   1.0 km

Fig. 21. Sketch showing a change of the coast-line occurred near Nevelsk as a result of the 2 August 2007 earthquake. a – position of the coast-line before the earthquake, b – uplifting portion of the sea-bottom after the earthquake; river Kazachka is shown as a black line.

4.4.5 Case 4 summary

In the case of the 2006 Gornozavodsk and the 2007 Nevelsk earthquakes the whole spectrum of prognoses from the long-term prediction to the short-term prediction of the 2007 Nevelsk earthquake was put into effect. The situation after the 2006 Gornozavodsk earthquake was interpreted correctly; the 2006 Gornozavodsk earthquake was treated as a foreshock of the stronger event. The short-term prediction was done for 7.5 months, but the 2007 Nevelsk earthquake had occurred three months later.

In case 4 the method of self-developing processes had resulted in an unexpectedly exact (Fig. 18) but maybe non-robust prognosis. The M8 algorithm was not applied in this case because of deficiency of data length for adjusting of the algorithm (the background level).

More details concerning the case 4 histories can be found in (Levin et al., 2007; Tikhonov, 2006; Tikhonov & Kim, 2010).

4.5 Case 5 - Unsuccessful intermediate-term prediction of a great earthquake at Southern Kuril Islands
4.5.1 Seismic region and data

This region includes the Urup, Iturup and Kunashir Islands (Fig. 22). The data used in earthquake forecasting were taken from the NEIC/USGS catalogues and contain earthquake

data from December 1995 until December 2007; events with M ≥ 4.0 were taken into account
as presumably registered without admissions.

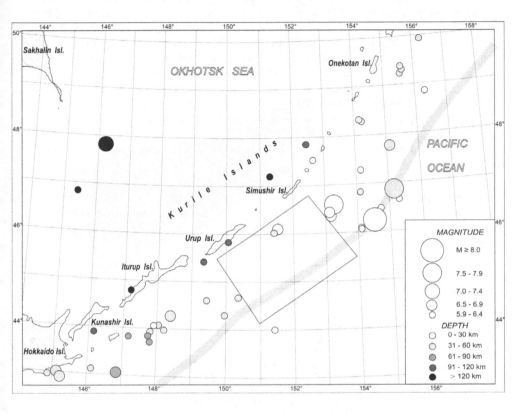

Fig. 22. Map of earthquake M ≥ 5.9 epicenters near Kuril Islands, May 1999 – January 2010.
The area limited by the polygon is the area of the seismic gap of the second kind.

4.5.2 Methodology

It was shown by the examples above that the M8 algorithm and the detection of seismic
gaps of the first and the second type provide a reasonable first approach to a long- and
intermediate-term earthquake prognosis. The decrease in b-values is known also as a
precursor of strong earthquake occurrence. In (Tikhonov, 1999, 2000) it was attempted to
present and apply a new formal algorithm for detection of both areas of intermediate- and
short-term seismic quiescence and change in b-value using a few functions that characterize
these features of seismic regime. This algorithm was named Q1. It was elaborated in analogy
with the structure of the M8 algorithm. The detailed description of the Q1 algorithm is
presented in (Tikhonov, 2000). We do not describe the Q1 algorithm here in detail because
the affectivity of application of this method was not supported by practice yet. For the
similar reasons we do not describe here the details of the case 5 history that can be found in
(Tikhonov, 2009).

4.5.3 Case 5 summary

The algorithm Q1 aimed at detection of joint occurrence of seismic gap and b-value change was presented in (Tikhonov, 1999; 2000). This algorithm has been applied to recognize the time of increased probability for large ($M \geq 7.5$) earthquake in the Southern Kuril Islands since the last large earthquake occurring here in December 03, 1995. Algorithm was examined at the time period since 1962 until 1995. Than the use of the algorithm in the real time regime has begun. The anomalies in four predictive functions of the Q1 algorithm were revealed on December 2007. In Fig. 22 the area of seismic quiescence statistically proved by the use of the Q1 algorithm is shown. The hazardous period for an earthquake $M \geq 7.4$ was declared for the next two years (2008-2009). Note that the use of the M8 algorithm (http://mitp.ru/predictions/html) had resulted in the alarm period for the similar space-time domain.

But the alarm has proved to be false. Till now there is no strong earthquake in the Southern Kuril Islands.

4.6 Case 1-5 summary

The summary of algorithms application is presented in Table 2. Note that the number of examples is insufficient for statistical estimation of relative validity of different applied algorithms.

Example No	Algorithms				
	M8	Seismic quiescence		Self-developing processes	Q1 algorithm
		ZMAP-technique	visual		
1	(+)	not used*	first type (not used) second type (+)	(+)	(no existed)
2	(+)	not used*	first type (not used) second type (+)	(not used)*	(no existed)
3	(-)	(+)	first type (not used) second type (not used)	(not used)	(no existed)
4	not used*	not used*	first type (+) second type (+)	(+)	(no existed)
5	(-)	not used*	first type (not used) second type (+)	(not used)	(-)

Table 2. The summary of algorithms application

(+) - means the algorithm was applied successfully;
(-) - means the algorithm failure;
(not used)* - means that the algorithm cannot be used due to different reasons;
(not used) - means the algorithm could be used but was not applied;
(not existed) – this new algorithm was developed later.

5. Discussion and conclusion

The analysis of behavior of seismicity within the generalized vicinity of large earthquake gives possibility to verify and to detail the characteristic parameters of the fore- and aftershock sequences and a number of other anomalies inherent to a vicinity of strong events (Rodkin, 2008). It was confirmed that the averaged fore- and aftershock cascades do obey the power law evolution. Power-law exponent of the foreshock cascade was found to be less than that of the aftershock cascade, and thus, the rate of increase of foreshocks number toward the moment of occurrence of the main event is slower than the rate of aftershocks decays. The typical duration of the aftershock process for M7+ events is about 100 days, while the average duration of the foreshock cascade in the constructed generalized vicinity was found to be quite noticeable during 10-20 days. The confirmation of a power law evolution for both fore- and aftershock cascades testifies that large earthquakes can be examined in terms of the critical phenomena. In this case it can be expected that the process of strong earthquake occurrence will be accompanied by other anomalies with a critical-like character of behavior. And factually, in parallel with the power-law fore- and aftershock cascades a stress-strain instability was shown to take place in the generalized vicinity of strong earthquake (see (Rodkin, 2008) for a more detailed description of these anomalies). It is worth mentioning also that much weaker increase in a number of events and the process of softening were revealed in a broader (few hundred days) time vicinity of a large earthquake beyond the domain of the fore- and aftershock cascades occurrence.

The set of precursory anomalies indicates the approaching of a strong event quite definitely. Thus one can conclude that the effective short- and intermediate-term earthquake forecasting appears to be possible in the case of an essential increase of volume of statistical information available for forecasting. Now in every particular case of earthquake forecasting the volume of available information is much less than it is available in the generalized vicinity of strong earthquake, and correspondingly the results of forecasting are expected to be substantially less certain. It does take place actually.

The state of art in a practice of earthquake forecasting is presented by an example of earthquake forecasting performed for the Sakhalin Island and the surrounding areas in the Institute of Marine Geology and Geophysics of the Far East Branch of the Russian Academy of Science, Yuzhno-Sakhalinsk, Russia. In two cases (1 and 4) from the five described above the whole set of earthquake parameters were successfully forecasted and thus these cases satisfy the term of "earthquake prediction". This practice suggests that at least in some cases earthquakes can be forecasted despite the shortage of available data.

All the used algorithms of earthquake forecasting are based upon the general properties of seismic regime in vicinity of strong earthquake. These properties (besides the seismic quiescence) are similar with those revealed in the generalized vicinity of strong earthquake. The "seismic quiescence" was not found in the generalized vicinity of strong earthquake because of anisotropic character of this type of precursor anomaly in relation to epicenter of the corresponding main shock.

We expect that the precursor features of the seismic regime behavior revealed in the generalized vicinity of strong earthquake can be useful in an earthquake prediction. These typical anomalies can be used as ideal images of precursory anomalies developing in process of preparation of individual strong earthquakes. Having in mind the volume of data used in the construction of the generalized vicinity of strong earthquake it can be suggested that a robust prognosis of strong earthquakes will be possible when the volume of data available in prognostic practice increases by one-two orders.

6. Acknowledgements

This work was supported by the Russian Foundation for Basic Research, grant No. 11-05-00663, and the European grant FP7 No. 262005 SEMEP.

7. References

Akimoto, T. & Aizawa, Y. (2006). Scaling Exponents of the Slow Relaxation in Non-hyperbolic Chaotic Dynamics. *Nonlinear phenomena in complex systems.* Vol.9, No.2, pp. 178-182, ISSN 1561-4085.

Bowman, D.D., Ouillon, G., Sammis, C.G., et al. (1998). An Observational Test of the Critical Earthquake Concept. *J. Geophys. Res.* Vol. 103, pp. 24359–24372, ISSN 0148-0227.

Geller, R.J. (1997). Earthquake prediction: a critical review. *Geophys. J. Inter.* Vol. 131, pp. 425–450, ISSN 1365-246X.

Geller, R.J., Jackson, D.D., Kagan, Y.Y. & Mulargia, F. (1997). Earthquakes cannot be predicted. *Science.* Vol. 275, pp. 1616–1619, ISSN 0036-8075.

Global Hypocenter Data Base CD-ROM. NEIC/USGS. - Denver, 1989.

Habermann, R. E. (1981). Precursory seismicity patterns: stalking the mature seismic gap. *Earthquake Prediction*, Maurice Ewing Series 4, D. W. Simpson & P. G. Richards, (Editors) American Geophysical Union, Washington, D.C., 2942.

Habermann, R. E. (1982). Consistency of teleseismic reporting since 1963. *Bull. of Seismol. Soc. Am.* Vol. 72. pp. 93-112, ISSN 0037-1106.

Habermann, R. E. (1983). Teleseismic detection in the Aleutian Islands arc. *J. Geophys. Res.* Vol. 88. pp. 5056-5064, ISSN 0148-0227.

Haken, H. (1978). *Synergetics.* Springer-Verlag, Berlin Heidelberg.

JMA Earthquake Catalog (Japan Meteorological Agency; 1926.1.1 – 2002.1.01).

Kagan, Y.Y. (1997). Are earthquakes predictable? *Geophys. J. Inter.* Vol. 131, pp. 505–525, ISSN 1365-246X.

Keilis-Borok, V.I. & Kossobokov, V.G. (1986). Time of Increased Probability for the Largest Earthquakes of the World. *Mathematical Methods in Seismology and Geodynamics, Comp. Seismol.* Vol. 19, pp. 48-57, ISSN 0203-9478, Nauka, Moscow (in Russian).

Keilis-Borok, V.I. & Kossobokov, V.G. (1990). Premonitory activation of earthquake flow: algorithm M8. *Physics of the Earth and Planetary Interiors.* Vol. 61, Nos. 1-2, pp. 73-83, ISSN 0031-9201.

Keilis-Borok, V.I. & Soloviev, A.A. (2003). *Nonlinear Dynamics of the Lithosphere.* Springer-Verlag, ISBN 354043528X, Berlin.

Keilis-Borok, V.I. & Soloviev, A.A., (Eds). (2002). *Nonlinear Dynamics of the Lithosphere and Earthquake Prediction.* Springer-Verlag, ISBN 978-3-540-43528-0, Berlin.

Keilis-Borok, V.I., Knopoff, L. & Rotwain, I.M. (1980). Bursts of aftershocks long term precursors of strong earthquakes. *Nature.* Vol. 283, pp. 259-263, ISSN 0028-0836.

Current State of Art in Earthquake Prediction, Typical Precursors and Experience in Earthquake Forecasting at Sakhalin Island and Surrounding Areas

75

Kim, Ch. U. (1989). Peculiarities of seismic energy release in space and time within the northern Sakhalin region. In: *The 1988 bulletin of Kuril-Sakhalin seismo-forecasting testing area (quarterly)*. No. 4. pp. 46-51, IMGG FEB RAS, Yuzhno-Sakhalinsk (in Russian).

Kossobokov, V.G. (2005). Earthquake Prediction: Principles, Implementation, Perspectives, In: *Earthquake Prediction and Geodinamic Processes, Comp. Seismol.* Vol. 36, pp. 1-179, ISSN 0203-9478, Nauka, Moscow (in Russian).

Kossobokov, V.G. (1986). Testing the algorithm M8: the Vrancha region. In: *Long-term earthquake prediction: methodical recommendations.* Sadovsky M.A. IPE AS USSR, Moscow, p. 102 (in Russian).

Kossobokov, V.G., Healy, J.H., Dewey, J.W., Shebalin, P.N. & Tikhonov, I.N. (1996). A real-time intermediate-term prediction of the October 4, 1994 and December 3, 1995 Southern-Kuril Islands earthquakes. *Comp. Seismol.* Vol. 28, pp. 46–55, ISSN 0203-9478, Nauka, Moscow (in Russian).

Kossobokov, V.G., Shebalin, P.N., Tikhonov, I.N., Healy, J.H. & Dewey, J.W., (1994). A real-time intermediate-term prediction of the October 4, 1994 Shikotan earthquake. In: *The Federal system of seismological observation and earthquake prediction. The informative-analytical bulletin. The 1994/10/4(5) Shikotan earthquake: Extraordinary issue.* Moscow, pp. 71-73 (in Russian).

Kraeva, N.V. (2003). Techniques and results of continued observations (1992-2002) of the South Sakhalin seismicity by the digital system IRIS. *Proceedings of problems of seismicity of the Far East and Eastern Siberia: Reports of the International Scientific Symposium,* Vol. 2, pp. 89-112, ISBN 5-7442-1358-9, Yuzhno-Sakhalinsk, September, 2002 (in Russian).

Latoussakis, J. & Kossobokov, V.G. (1990). Intermediate Term Earthquake Prediction in the Area of Greece: Application of the Algorithm M8. *Pure Appl. Geophys.* Vol. 134, No. 2, pp. 261-282, ISSN 0033-4553.

Lennartz, S., Bunde, A. & Turcotte, D.L. (2008). Missing data in aftershock sequences: Explaning the deviations from scaling laws. *Rev. E. Stat. Nonlin Soft Matter Phys,* Vol. 78, pp. 41115-41123, ISSN 1550-2376.

Levin, B.V., Kim, Ch.U., Tikhonov, I.N. (2007). The Gornozavodsk earthquake of 17(18) August, 2006, in the south of Sakhalin Island. *J. Pacific Geol.* Vol. 1, No. 2, pp. 102–108, ISSN 0207-4028 (in Russian).

Lindman M., Lund, B. & Roberts, R. (2010). Spatiotemporal characteristics of aftershock sequences in the South Iceland Seismic Zone: interpretation in terms of pore pressure diffusion and poroelasticity. *Geophys. J. Int.* Vol. 183, No. 3, pp. 1104–1118, ISSN 1365-246X.

Malamud, B.D., Morein, G. & Turcotte, D.L. (2005). Log-periodic behavior in a forest-fire model. *Nonlinear Processes in Geophysics.* Vol. 12, pp. 575-585, ISSN 1023-5809.

Malyshev, A.I. (1991). Dynamics of self-developing processes. *J. Volcanology and Seismology.* No. 4, pp. 61–72, , ISSN 0203-0306 (in Russian).

Malyshev, A.I., Tikhonov, I.N. & Dugartsyrenov, K.Ts. (1992). The technique of mathematical modeling the Kurile foreshock-aftershock strong earthquake sequences. Preprint IMGG, Yuzhno-Sakhalinsk (in Russian).

Mogi, K. (1985). *Earthquake prediction*. Academic Press (Harcourt Brace Jovanovich, Publishers), New York.

Oskorbin, L.S. & Bobkov, A.O. (1997). Seismic behavior of the Far East seismogenic zones. In: *Problems of seismic hazard of Far East region: Geodynamic of tectonosphere of the Pacific–Eurasia conjunction zone*, Tarakanov R.Z. & Ivaschenko A.I., Vol. VI, pp. 179–197, IMG&G, ISBN 5-7442-1028-8 (T. 6),Yuzhno–Sakhalinsk (in Russian).

Papazachos, C.B., Karakaisis, G.F., Scordilis, E.M. & Papazachos, B.C. (2005). Global Observational Properties of the Critical Earthquake Model. *Bull. Seismol. Soc. Am.* Vol. 95, No. 5, pp. 1841-1855, ISSN 0037-1106.

Quick Epicenter Determination (QED). The NEIC/USGS Branch of Global Seismology and Geomagnetism On-line Information System, 1992.

Rodkin, M.V. (2008). Seismicity in the Generalized Vicinity of Large Earthquakes. *J. Volcanology and Seismology*. Vol. 2, No. 6, pp. 435–445, ISSN 0203-0306 (in Russian).

Romashkova, L.L. & Kosobokov, V.G. (2001). The Dynamics of Seismic Activity before and after Great Earthquakes of the World, 1985–2000. *Comp. Seismol.* Vol. 32, pp. 162–189, ISSN 0203-9478, Nauka, Moscow (in Russian).

Shebalin, N.V. (1961). Intensity, magnitude and depth of an earthquake source. *Earthquakes in USSR*. AS USSR, Moscow, pp. 126–138 (in Russian).

Shebalin, P.N., (2006). A Methodology for Prediction of Large Earthquakes with Waiting Times Less than One Year. *Comp. Seismol.* Vol. 37, pp. 7–182, ISSN 0203-9478, Nauka, Moscow (in Russian).

Shimamoto, T., Watanabe, M., Suzuki, Y., Kozhurin, A.I., Streltsov, M.I. & Rogozhin, E.A. (1996). Surface faults and damage associated with the 1995 Neftegorsk earthquake. *J. Geol. Soc. Jpn.* Vol. 102 (10), pp. 894–907, ISSN 1684-9876.

Smirnov, V.B. & Ponomarev, A.V. (2004). Patterns in the Relaxation of Seismicity from Field and Laboratory Observations. *Izv. RAN, Fizika Zemli*. No. 10, pp. 26–36, ISSN 0002-3513 (in Russian).

Smirnov, V.B., (2003). Estimating the Duration of the Fracture Cycle in the Earth's Lithosphere from Earthquake Catalogs. *Izv. RAN, Fizika Zemli*. No. 10, pp. 13–32, ISSN 0002-3513 (in Russian).

Sobolev G.A., Tyupkin Yu.S. & Zavyalov A.D. (1999). Map of expected algorithm and RTL prognostic parameter: joint application. *Russ. J. Earthquake Sciences*. Vol. 1. No 4. pp. 301-309, ISSN 1681-1206.

Sobolev, G.A. & Ponomarev, A.V. (2003). *Physics of earthquakes and precursors*. Nauka, ISBN 5-02-002832-0, Moscow (in Russian).

Sobolev, G.A. (1993). *Principles of Earthquake Prediction*. Nauka, ISBN 5-02-002287-X, Moscow (in Russian).

Sornette, D. (2000). *Critical Phenomena in Natural Sciences*. Springer-Verlag, ISBN 354067424 , Berlin–Heidelberg.

Streltsov, M.I. (2005). *The May 27(28), 1995 Neftegorsk earthquake on Sakhalin Island.* Ivaschenko A.I., Kozhurin A.I. & Levin B.W. Yanus-K, ISBN 5-8037-0256-0, Moscow (in Russian).

Tanaka, K. (1980). Formation pattern of seismic gaps before and after large earthquakes. *Zisin. J. Seismol. Soc. Jpn.* Vol. 33, No. 3, pp. 369–377.

Tarakanov, R.Z. (1995). Source dimensions of large Kuril–Kamchatka and Japan earthquakes and maximum possible magnitude problem. *J. Volcanology and Seismology.* No. 1, pp. 76–89, ISSN 0203-0306 (in Russian).

Tikhonov, I. N. (1999). A method of intermediate-term prediction of time occurrence of strong (M ≥ 7.5) earthquakes (on the example of the territory around the Southern Kurile Islands), Preprint IMGG, Yuzhno-Sakhalinsk (in Russian).

Tikhonov, I. N. (2001). A method of intermediate-term prediction of probably periods of occurrence of strong earthquakes in application to the Kuril Islands region. *Proceedings of problems of geodynamics and earthquakes forecasting. The I Russian-Japanese Workshop,* pp. 158-169, ISBN 5-7442-1275-2, Khabarovsk, September, 2000 (in Russian).

Tikhonov, I.N. (1997). Some patterns in seismic regime dynamics of the Southern Sakhalin region. *Bull. Seismol. Assoc. Far East,* Vol. 3 No. 2, pp. 192–211.

Tikhonov, I.N. (2000). Precursors of the 1995 Neftegorsk earthquake and a recent precursory situation in the southern Sakhalin, *Proceedings of a memory and lessons of the 1995 Neftegorsk earthquake. The scientific-technical seminar-meeting. Collected reports,* pp.72-74, ISBN 5-94137-015-7, Yuzhno-Sakhalinsk, May 2000. POLTEX, Moscow (in Russian).

Tikhonov, I.N. (2003). Seismic quiescence before the strong earthquakes of Japan, *Proceedings of XXIII General Assembly of the International Union of Geodesy and Geophysics,* Sapporo, Japan, June – July, 2003. Abstracts Week A, P A.479-A.480.

Tikhonov, I.N. (2005). Detection and mapping of seismicity quiescence prior to large Japanese earthquakes. *J. Volcanology and Seismology.* No. 5, pp. 1–17 (in Russian).

Tikhonov, I.N. (2006). *Methods of earthquake catalog analysis for purposes of intermediate- and short-term prediction of large seismic events.* Vladivostok, Yuzhno–Sakhalinsk: IMGG FEB RAS, ISBN 5-7442-1415-1, Yuzhno–Sakhalinsk (in Russian).

Tikhonov, I.N. (2009). A technique of the strong earthquake prediction from the flux of seismicity in the North-Western part of the Pacific belt. Ph. Dr. Thesis. IMGG FEB RAS, Yuzhno-Sakhalinsk (in Russian)

Tikhonov, I.N., Kim, Ch.U. (2010). Confirmed prediction of the 2 August 2007 MW 6.2 Nevelsk earthquake (Sakhalin Island, Russia). *Tectonophysics.* Vol. 485, issues 1-4, pp. 85-93, ISSN 0040-1951.

Utsu, T., Ogata, Y., & Matsu'ura, R.S. (1995). The Century of the Omori Formula for Decay Law of Aftershock Activity. *J. Phys. Earth,* Vol. 43, pp. 1–33.

Vvedenskaya N.A., Kondorskaya N.V. et al. (Eds.). 1964 – 1991. *Earthquakes in USSR, 1962 – 1990,* Nauka, Moscow (in Russian).

Wiemer, S., Wyss, M. (1994). Seismic quiescence before the Landers (M=7.5) and Big Bear (M=6.5) 1992 earthquakes. *Bull. of Seismol. Soc. Am.* Vol. 84. No. 3, pp. 900-916, ISSN 0037-1106.

Zavyalov, A.D. (2006). *Intermediate-Term Earthquake Prediction: Principles, Techniques, Implementation.* Nauka, ISBN 5-02-033946-6, Moscow (in Russian).

4

Earthquakes Precursors

Dumitru Stanica and Dragos Armand Stanica

Institute of Geodynamics of the Romanian Academy
Romania

1. Introduction

Strong earthquake of magnitude 7 or more (on the Richter scale) strikes about once a year somewhere in the world and, several times triggers a cascade of follow-on events, such as tsunamis, floods, landslides, nuclear power plant crisis and public health catastrophes in the affected regions. Thus, during the 2004 Sumatra–Andaman earthquake and Indian Ocean tsunami nearly 230,000 people were killed and more than one million people were left homeless in 13 countries surrounding the Indian Ocean. The May 12th, 2008 earthquake in Western Sichuan, China and January 8th, 2010 earthquake in Haiti caused a death toll well over 75,000 and 320,000 people, respectively. The latest M9 Tohoku earthquake of March 11th 2011 in Japan was the biggest recorded earthquake ever to hit Japan. The earthquake triggered extremely destructive tsunami waves of up to 10 meters that struck Japan minutes after the quakes and caused about 26,000 deaths and 3000 injured. Recent catastrophic earthquakes (2004–2011) occurred in Asia, Europe and America have provided and renewed interest in question of the existence of precursory signals related to earthquakes. In these circumstances, the science community is struggling on how to provide early information related to the occurrence time of such events in order to reduce the loss of human life and property. Previous studies (Gotoh et al., 2002; Fraser-Smith et al., 1990; Freund et al., 1999; Hattori et al., 2006; Hayakawa & Fujinawa, 1994; Hayakawa & Molchanov, 2002; Kopytenko et al., 1994; Liu et al., 2004; Ouzounov et al., 2006; Parrot et al., 2007; Pulinets et al., 2004; Stanica & M. Stanica, 2007; Stanica & D.A. Stanica, 2010; Tramutoli et al., 2005; Tronin et al., 2004; Varotsos, 2005) have shown that there were precursory signals observed on the ground and in space associated with several earthquakes. In the last 10 years, the interdisciplinary group for Electromagnetic Study of Earthquakes and Volcanoes (EMSEV) have demonstrated that the existence of the electromagnetic earthquake precursors by terrestrial and satellite observations is not trivial, and it is necessary a wide international cooperation and several more years of research with primary focus in the following directions: (i) what is the possible generation mechanisms of the electromagnetic phenomena; and (ii) whether electromagnetic precursors systematically precede earthquakes. In this respect, taking into account that the seismic-active Vrancea zone, Romania is one of the "hot" subjects in the Eastern Europe, this paper is focused on the specific methodology able to emphasize the short-term electromagnetic (EM) precursory parameters, associated to intermediate depth earthquakes (70-180Km). We consider that one of the realistic mechanisms for triggering such events in the seismogenic volume can be the dehydration of rocks which make fluid-assisted faulting possible. The changes of electrical conductivity occurred before an earthquake, as a sequence of geodynamic processes associated with fluid migration through faulting system developed into and in the close vicinity of the seismogenic volume, could be detected by means of the anomalous behavior of the Bzn parameter taken throughout the frequency range less than 1.66E-2 Hz (Stanica & M. Stanica, 2007; Stanica & D.A. Stanica, 2010). According to the electromagnetic information acquired in 2009-2010 years correlated with seismic events, it is relieved that some days before an EQ occurred, the daily mean variation of the Bzn parameter have an anomalous behavior marked by a significant increase versus its normal distribution identified in non seismic conditions.

2. Geodynamic models and possible mechanisms of intermediate depth earthquakes

The seismic active Vrancea zone is situated at the curvature of the Carpathians and it is bounded on the north-east by the East European Platform, to southward by the Moesian Platform, and on the north-west by Transylvanian Basin (Fig.1).
The hypocenters of the intermediate depth earthquakes are concentrated within a very small seismogenic volume and they are much denser than any other mantle events of intra-continental origin known in the world. In the past century, 4 large seismic events have

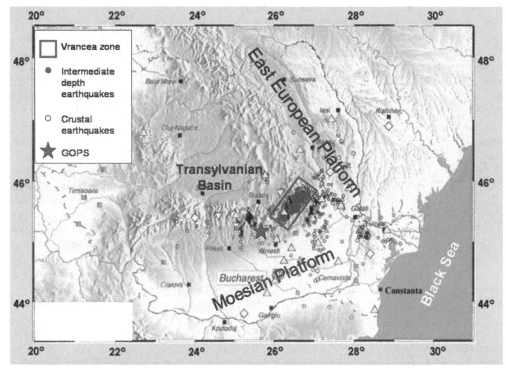

Fig. 1. Map of the seismic active Vrancea zone: crustal (white circles) and intermediate depth (red circles) epicenters of the earthquakes taken from ROMPLUS catalogue; pink star represents the Geodynamic Observatory Provita de Sus (GOPS) used as monitoring site of the earthquake precursors

occurred in the intermediate depth range of 70 to 180 km (in 1940 with moment magnitude Mw7.7, in 1977 with Mw7.4, in 1986 with Mw7.1, and in 1990 with Mw 6.9) and all of them cause destruction in Bucharest, the capital city of Romania, and shake central and eastern European cities several hundred kilometers away from their epicenters.

Several geodynamic models related to the triggering mechanism of the intermediate depth earthquakes have been elaborated in this area. Oncescu (1984) and Oncescu at al., (1984) proposed a double subduction model on the basis of 3-D seismic tomographic images: in their interpretation, the intermediate-depth earthquakes are generated within a vertical surface separating the sinking slab from stable lithosphere.

Trifu & Radulian (1989), analyzing the seismic behavior of the Vrancea zone, proposed a model based on the existence of two active zones located at depths of 80-110 km and 120-170 km. Both zones are characterized by local stress inhomogeneities capable of generating large earthquakes.

Khain & Lobkosky (1994) suggest that the Vrancea zone results from delamination processes occurred during continental collision and lithosphere sinking into the mantle.

Linzer (1996) explains the nearly vertical position of the Vrancea slab as the final rollback stage of a small fragment of oceanic crust.

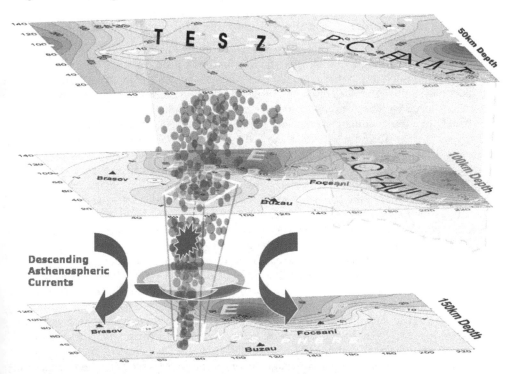

Fig. 2. 3D resistivity tomographic image at sub-crustal level in the seismic active Vrancea zone: red circles are intermediate depth earthquakes; blue square delineate the Trans-European Suture Zone (M. Stanica et al., 1999); green line is Peceneaga-Camena fault (P-C fault); pink arrows show the direction of the asthenospheric currents; red arrow shows the direction of the torsion process of the relic slab

Sperner, B., the Collaborative Research Center [CRC] 461 Team, (2005), taking into consideration that the geometry of the subduction zone was not unequivocally defined, proposed four possible configurations for the Vrancea zone: (i) subduction beneath the suture zone; (ii) subduction beneath the fore deep area; (iii) two interacting subduction zones, and (iv) subduction beneath the suture, followed by delamination.

Various types of slab detachment or delamination have been proposed to explain the present-day seismic images of the descending slab (Girbacea & Frisch, 1998; Gvirtzman, 2002; Sperner et al., 2001; Wortel & Spakman, 2000).

Viscous flows due to the sinking seismogenic slab together with dehydration-induced faulting can be considered as possible triggering mechanism explaining the intermediate-depth seismicity in Vrancea (Ismail-Zadeh et al., 2000)

Stanica et al., (2004) show, on the base of the three-dimensional (3D) resistivity tomographic image carried out using magnetotelluric data, that the possible triggering mechanism of the intermediate-depth earthquakes in the Vrancea zone may be the rock response to the active torsion processes sustained by the descending asthenospheric currents and the irregular shape of the relic slab. In their opinion, this torque effect may generate the increase shear stress and drive faulting process within the rigid slab (Fig.2).

3. Generation mechanisms and theoretical base of the electromagnetic precursors to earthquakes

3.1 Possible generation mechanisms of electromagnetic precursors

This paper is not intended to present an exhaustive analysis of all of literature published in this field. Instead, we have tried to provide only some representative hypothetic mechanisms related to the electromagnetic precursors to earthquakes.

The theory of semiconductors launched by Freund (2000) is considered the most comprehensive one. According to this theory, the solid rocks (e.g. granites) begin to crack under the action of a stress exceeding their elasticity limit, what leads to a release of electric charges. These charges, carried by water moving through the rocks fissures, generate currents of high amperage that, in their turn, create disturbances of the magnetic field and, also, infrared signature in the bands of 8μm and 11μm, when the charges are neutralized at the surface of the Earth.

In accordance with this theory, to have seismo-electric signals it is necessary that a few compulsory conditions to be fulfilled:

- Presence of major tectonic accidents able to produce an important stress;
- Existence of quartz or rocks with a rich content of quartz;
- Fluid carrying electric charges through porous rock what generates electric currents that lead to disturbances of the magnetic field.

The resistivity of porous rocks is changed as a function of compression and shearing (Brace et al., 1965) and may be measured by using passive experiments (magnetotelluric method), or experiments with active methods (Park et al., 1993).

Revol et al., (1977) have shown that the magnetic properties of rocks are changed depending on the applied stress and are associated to the changes of stress emphasized on fault rupture, what produces oscillations of the magnetic field of a few nT (Johnston, 1978, 1997); this mechanism is due to the piezomagnetic effect.

When a conductive fluid is forced to flow close to a surface with stationary electric charges, an electrokinetic effect appears generating currents that start to flow either through fluid, or through the surrounding rock, what, in conditions imposed by real crustal parameters, may create surface magnetic fields of a few nT (Fenoglio et al., 1995).

Another theory supposes the generation of the magnetic signal either by conductive fluid flowing in presence of the magnetic field of the Earth, or by magnetohydrodynamic conversion of the seismic signal into an electric signal during the propagation through a conductive medium (Molchanov et al., 2001). While these mechanisms were proved in laboratory conditions, it is unclear, yet, how this process takes place in conditions rather similar to those specific to the Earth, owing to the lack of measurements in active fault zones.

3.2 Theoretical base of the electromagnetic precursors

At the Earth surface the vertical geomagnetic component (Bz) is entirely secondary field and its existence is an immediate indicator of lateral inhomogeneity. For a two-dimensional (2D) structure, the vertical geomagnetic component (Bz) is produced essentially by the horizontal geomagnetic component perpendicular ($B_\perp$) to geoelectric structure orientation and, consequently, the normalized Bzn function defined as:

$$Bzn\,(f) = \frac{Bz(f)}{B_\perp(f)} \tag{1}$$

should be time invariant in non geodynamic conditions (Ward et al., 1970), but it becomes unstable due to the geodynamic processes and, therefore, it could be used as a precursory parameter of the intermediate depth seismic activity (Stanica and D.A. Stanica, 2010).

In order to explain cause (earthquake) - effect (anomalous Bzn) relationship, we introduce the following equations:

$$\rho z\,(f) = \frac{0.2}{f}\left|\frac{E_{\parallel}(f)}{Bz(f)}\right|^2 \tag{2}$$

where: ρz is vertical resistivity [$\Omega m = VmA^{-1}$], f is the frequency [Hz] and the $E_{\parallel}$ is the electric field parallel to strike [Vm^{-1}], Bz is the vertical component of the magnetic induction [Tesla (T) = V s m^{-2}].

Also, it is possible to write the relation:

$$\rho_{\parallel}(f) = \frac{0.2}{f}\left|\frac{E_{\parallel}(f)}{B_\perp(f)}\right|^2 , \tag{3}$$

where: $\rho_{\parallel}$ is the resistivity parallel to strike[Ωm], $B_\perp$ is the component of the magnetic induction perpendicular to strike
[Tesla (T) = V s m^{-2}].

From the relations (2) and (3) we may estimate the normalized function Bzn (f), in terms of resistivities as follows:

$$\left|Bzn\,(f)\right| = \sqrt{\frac{\rho_{\parallel}(f)}{\rho z(f)}} \tag{4}$$

This estimation of Bzn is in error for non two-dimensional geoelectrical structure.

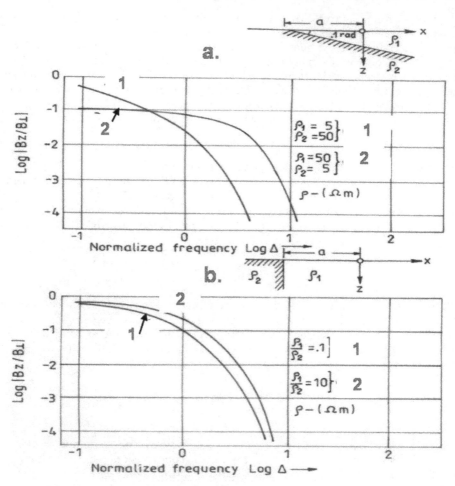

Fig. 3. Bzn distribution versus normalized frequency for sloping interface (a.) and vertical contact (b.): $\Delta = a/\delta_1$

The relation (4) demonstrates that normalized function Bzn could be linked to the resistivity/conductivity variation along the faulting systems acting as high sensitive path (represented by the Carpathian electrical conductivity anomaly) through the lithosphere and its right part lead to the normalized resistivity defined as:

$$\rho n(f) = \frac{\rho_{\parallel}(f)}{\rho z(f)} \tag{5}$$

Approximate field solutions were computed for two simple 2D geoelectric structures to illustrate the robustness of the relation (1). Solutions for the sloping interface and vertical contact models were obtained using finite element code (Wannamaker et al., 1986) and the results are presented in Fig. 3. These models represent two extremes in dipping angle of the interface and similarity in the properties of the normalized function Bzn obtained for the model (Fig.3, b.) is of interest in selecting the site (GOPS) for continuous monitoring of the geomagnetic field. The normalized frequency scale (Δ) is proportional to distance along the x-axis and inverse-proportional to penetration depth (δ_1) in medium of resistivity ρ_1. The electromagnetic skin depth or penetration depth is given by:

$$\delta_1[km] = \frac{1}{2\pi}\sqrt{\frac{10\rho_1[\Omega m]}{f[Hz]}} \qquad (6)$$

4. Electromagnetic (EM) methodology and results

4.1 Electromagnetic data collection

As we have seen in relation (1), Bzn could be used as precursory parameter of seismic event by measuring the vertical geomagnetic component (Bz) and horizontal component perpendicular to the strike ($B_\perp$) which have been collected at the Geodynamic Observatory Provita de Sus (GOPS), placed on the Carpathian electrical conductivity anomaly (CECA). This anomaly is delineated by the Wiese induction arrows, and it can represent a zone of partial melting or of hot highly-mineralized fluids in sedimentary layers, formed at the collisional limit between the both platforms (East European and Moesian) with Carpathian Alpine structures (Fig.4). It is also quit possible that these two varieties of fluid anomalies to co-exist and gradually flow one into another, as indicated by the fact that geoelectric parameters remain fairly constant throughout its entire length (Pinna et al., 1993, Rokityansky & Ingerov, 1999).

Induction arrows are vector representations of the ratio of vertical to horizontal magnetic field components. Since vertical magnetic fields are generated by lateral conductivity gradients, induction arrows map can be used to infer the presence, or absence of lateral variation of conductivity/resistivity. In the Wiese convection (Wiese, 1962) the vectors point away from the conductivity anomaly generated by anomalous internal concentrations of current, while in the Parkinson convection (Parkinson, 1959), the vectors point towards anomalous internal concentrations of current. Thus, insulator-conductor boundaries extended through a 2D geoelectrical structure (like CECA) give rise to induction arrows that orientate perpendicular to their geoelectrical strike, and have magnitude proportional to the intensity of anomalous current concentration (Jones & Price, 1970), which are in turn determined by the magnitude of conductivity gradient.

In our methodology, it was also supposed that pre-seismic conductivity changes, due to the fluid migration through faulting system, may generate changes of the normalized function Bzn, having magnitude proportional to the intensity of anomalous current concentrations through CECA.

The Geodynamic Observatory Provita de Sus (Fig.1) is located at about 100 km towards south-west of seismic active Vrancea zone and the criteria of selection as monitoring site are:

- Existence of logistic base able to supply optimal EM data;
- Placement on the Carpathian electrical conductivity anomaly where, ideally, the condition for a 2D type geoelectric structure is fulfilled (Fig. 4);
- Real time wireless data transfer to the central office (Institute of Geodynamics, Bucharest).

In order to select the frequency range where the relation (1) is valid (i.e., existence of a 2D geoelectrical structure and its strike orientation), as a first step in our EM methodology, at the GOPS we made a magnetotelluric sounding using the magnetotelluric (MT) equipment

GMS-06 (METRONIX - Germany). This geophysical system has 5 channels (two electric Ex, Ey and three magnetic Bx, By, Bz components), 24 bit resolution, GPS, two frequency ranges (LF: 4096sec.-1kH; HF=0.5kH-10kH) and for data processing "MAPROS" software packages.

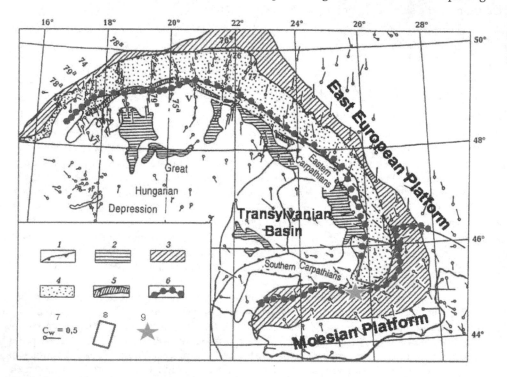

Fig. 4. Carpathian electrical conductivity anomaly and Wiese induction arrows on a tectonic sketch map: 1) main boundaries and fractures (over thrusts) of regional structures; 2) Neogene volcanic rocks; 3) Carpathian fore-deep; 4) Carpathian flysch nape system; 5) Piena and Marmarosh cliff belt; 6) Carpathian electrical conductivity anomaly (CECA); 7) Wiese induction vectors magnitude; 8) seismic active Vrancea zone (intermediate-depth earthquakes); 9) Geodynamic Observatory Provita de Sus (GOPS) used for the electromagnetic data collection (Modified after Rokityansky & Ingerov, 1999)

It is well known that the magnetotelluric (MT) method is a passive technique that involves measuring fluctuations of natural electric (E) and magnetic (B) fields in orthogonal directions at the surface of the Earth (Kaufman, & Keller, 1981). The orthogonal components of the horizontal electric (Ex, Ey) and magnetic (Bx, By) fields are related by the complex impedance tensor, Z:

$$\begin{pmatrix} Ex \\ Ey \end{pmatrix} = \begin{pmatrix} Zxx & Zxy \\ Zyx & Zyy \end{pmatrix} \begin{pmatrix} Bx \\ By \end{pmatrix}, \text{ or } E = ZB \tag{7}$$

Where: Zxx, Zxy, Zyx, Zyy are elements of the impedance tensor [VA⁻¹]
For a 2D structure, in which the conductivity varies along one horizontal direction as well as with depth, the following relations are fulfilled:

$$Zxx = -Zyy$$
$$Zxy \neq -Zyx$$

(8)

Using single site magnetotelluric impedance tensor decomposition technique (Bahr, 1988), it was possible to identify the following two parameters: skewness and strike orientation. The skewness is a dimensionality parameter of the impedance tensor, defined as:

$$Skew = \frac{\|Zxx + Zyy\|}{\|Zxy + Zyx\|}$$

(9)

This parameter should be less than 0.3 to interpret the structure as 2D.
The tensor impedance from relation (7) can be rotated to obtain the strike orientation of the 2D geoelectrical structure using the relation:

$$\alpha = \frac{1}{4}\arctan\frac{2Re(Zxy + Zyx)(\overline{Zxx - Zyy})}{(|Zxx - Zyy|^2 - |Zxx + Zyx|^2)}$$

(10)

Where: α (strike) is rotation angle [^{0}].

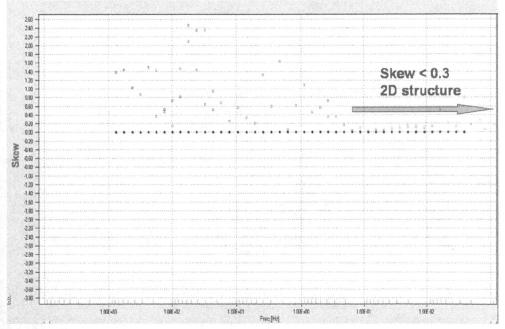

Fig. 5. Skewness parameter versus frequency (Hz): pink arrow delineates the frequency range for 2D structure

The MAPROS software packages include all these mathematical relations presented above and have been applied to MT sounding carried out at GOPS. This program performs the following basic tasks:

- Real time data acquisition and processing;
- Robust estimation of the magnetotelluric transfer functions;
- Real time display of time series and all important electromagnetic parameters ($\rho_\perp$, $\rho_\parallel$, skewness and strike, etc).

Thus, on the base of MT results, a 2D geoelectrical structure has been identified on the frequency range less than 1.66 E-2 Hz where skewness < 0.3 (Fig.5) and average strike orientation is N96^{0}E (Fig.6). This frequency range is also associated with the intermediate-depth earthquakes interval (70-180km) where EM precursors are generated.

These results confirm, once more, that the CECA's geoelectrical structure is of 2D type with strike orientation approximately east-west, and forms not only a tectonic boundary between Moesian Platform and Carpathian Alpine structures, but also represents a peculiar conducting channel extended to the seismic active Vrancea zone (Fig.4).

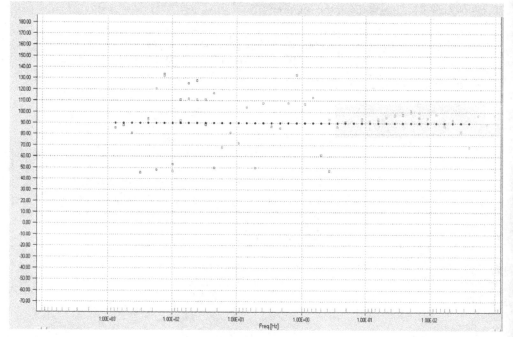

Fig. 6. Strike orientation (degrees) versus frequency (Hz): blue rectangle delineates the frequency range for 2D structure with average strike orientation of N96^{0}E

The next step in our study was to realize a continuous monitoring of the geomagnetic components ($B_\perp$, $B_\parallel$, Bz) using the acquisition module MAG-03 DAM (Bartington-England), with 6 channels, 24 bit resolution and three axis magnetic field sensor MAG-03 MSL (frequency range: DC - 1kHz). In order to obtain $B_\perp$ component of the geomagnetic field, one of the horizontal components of the three axis magnetic sensor must be orientated perpendicular to strike. The parameters of the data acquisition card are under software control and additional program collects information at each five seconds and stored them, every 60 seconds (Table.1), on the PC HD. Using the wireless connection, all the data are transferred from GOPS to the central unit, placed at the Institute of Geodynamics in Bucharest, for real-time data processing and analysis (Fig.7).

Fig. 7. Monitoring system of the geomagnetic components (Bx, By and Bz) and real time data transfer : acquisition module MAG-03DAM (a); computer (b) for data storage; monitor (c) for real time geomagnetic data display (d); data transfer program (e) ; wireless connection (f)

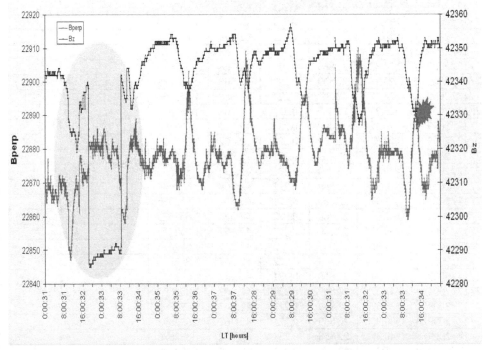

Fig. 8. Geomagnetic time series (Bperp and Bz) recorded at the GOPS for 7 days interval (April19- April 25, 2009); Bperp is $B_\perp$; red star is earthquake of M5.0; pink ellipse marks a pre-seismic disturbance of the vertical component (Bz) of geomagnetic field (lead time is about 6 days before the earthquake).

Date	Time [s]	$B_\perp$[μT]	$B_\parallel$[μT]	Bz[μT]	Bzn	Bzn average (1day data)	STDEV
7/17/2009	0:00:37	22.869	0.22	42.367	1.852595	1.852561	0.000281
7/17/2009	0:01:37	22.869	0.22	42.367	1.852595		
7/17/2009	0:02:37	22.869	0.22	42.367	1.852595		
7/17/2009	0:03:37	22.868	0.22	42.367	1.852676		
7/17/2009	0:04:37	22.868	0.22	42.367	1.852676		
7/17/2009	0:05:37	22.868	0.221	42.367	1.852676		
7/17/2009	0:06:37	22.869	0.22	42.367	1.852595		
7/17/2009	0:07:37	22.868	0.22	42.367	1.852676		
7/17/2009	0:08:37	22.868	0.221	42.367	1.852676		
7/17/2009	0:09:37	22.868	0.22	42.367	1.852676		
7/17/2009	0:10:37	22.868	0.22	42.367	1.852676		
7/17/2009	0:11:37	22.869	0.221	42.368	1.852639		
7/17/2009	0:12:37	22.868	0.22	42.368	1.85272		
7/17/2009	0:13:37	22.868	0.221	42.368	1.85272		
7/17/2009	0:14:37	22.869	0.221	42.368	1.852639		
7/17/2009	0:15:37	22.868	0.22	42.367	1.852676		
7/17/2009	0:16:37	22.868	0.221	42.367	1.852676		
7/17/2009	0:17:37	22.869	0.221	42.367	1.852595		
7/17/2009	0:18:37	22.868	0.221	42.367	1.852676		
7/17/2009	0:19:37	22.869	0.221	42.367	1.852595		
7/17/2009	0:20:37	22.869	0.221	42.368	1.852639		
7/17/2009	0:21:37	22.869	0.221	42.368	1.852639		
7/17/2009	0:22:37	22.869	0.221	42.368	1.852639		
7/17/2009	0:23:37	22.869	0.221	42.368	1.852639		
7/17/2009	0:24:37	22.87	0.221	42.368	1.852558		
7/17/2009	0:25:37	22.869	0.221	42.368	1.852639		
7/17/2009	0:26:37	22.869	0.221	42.367	1.852595		
7/17/2009	0:27:37	22.869	0.221	42.367	1.852595		
7/17/2009	0:28:37	22.869	0.221	42.367	1.852595		
7/17/2009	0:29:37	22.869	0.221	42.367	1.852595		
7/17/2009	0:30:37	22.869	0.221	42.368	1.852639		
7/17/2009	0:31:37	22.869	0.221	42.367	1.852595		
7/17/2009	0:32:37	22.869	0.222	42.367	1.852595		
7/17/2009	0:33:37	22.869	0.222	42.367	1.852595		
7/17/2009	0:34:37	22.868	0.222	42.367	1.852676		
7/17/2009	0:35:37	22.869	0.222	42.367	1.852595		
7/17/2009	0:36:37	22.868	0.221	42.367	1.852676		
7/17/2009	0:37:37	22.869	0.221	42.367	1.852595		
7/17/2009	0:38:37	22.869	0.221	42.367	1.852595		
7/17/2009	0:39:37	22.869	0.221	42.367	1.852595		
7/17/2009	0:40:37	22.868	0.221	42.367	1.852676		
7/17/2009	0:41:37	22.869	0.221	42.367	1.852595		
7/17/2009	0:42:37	22.868	0.222	42.367	1.852676		

Table 1. Geomagnetic time series $B_\perp$, $B_\parallel$ and Bz recorded on July 17th 2009 (42 minutes record): Bzn average is computed for 1day data; STDEV of the Bzn average.

The geomagnetic time series recorded at the GOPS for 7 days interval (April19- April 25, 2009), including the occurrence time of the earthquake of M5.0 (April 25), are presented in Fig. 8. The pre-seismic disturbance of the vertical component (Bz) occurred about 6 days before earthquake. But, as we have seen later on, this disturbance is masked by the superposition effect started on March 9, 2009.

4.2 Results

In this paper, daily mean distribution of the normalized function Bzn and its standard deviation are performed in the frequency range less than 1.666E-2 Hz, where 2D structural condition is fulfilled. The concept of this analysis is based on the idea that signal associated with solar-terrestrial origin is constant, according to relation (1), while lithospheric origin signal from the underground current flowing along the CECA is considered to have a vertical component (see fig.8). With the other words, the normalized function Bzn shows a small and certain value for its normal trend (in non seismic condition) and increased values in pre-seismic conditions.

To assess the robustness of the presented methodology, some examples of Bzn distribution acquired in a span of about two years (2009 -2010) are shown in correlation with the intermediate depth earthquakes, with magnitude (Mw) higher than 4.0 (Richter scale), selected from the catalogue issued by National Institute of the Earth Physics-Bucharest.

The first particular case of the Bzn distribution correlated with the both standard deviation (STDEV) and intermediate depth earthquakes, within the interval January 16 – May 11, 2009 is shown in Fig. 9.

The Bzn distribution emphasizes two domains, the first one, with normal values of about 1.842 on the interval January16 - March 8 and second one, on March 9- May 11 interval, having values between 1.850-1.856, and all earthquakes are marked by vertical arrows.

Average value of 1.842, associated with earthquakes of M< 3.3 occurred on the interval January 16 – March 8 represents the threshold limit between the so called "normal trend" of Bzn and its second anomalous domain, which started on March 9, which may represent a superposition effect of the four earthquake of M4.0 (March 21), M4.1 (April 12), M5.0 (April 25) and M5.0 (May 11).

The earthquake of magnitude 5.0 was triggered in the Vrancea zone, at 109 km depth, on April 25 at 20:18:48 (local time), being felt in Bucharest and over a large area extended from the epicentral zone towards NE and SW directions, corresponding with the fault plane orientation of the focal mechanism.

Similar results have been obtained in the Bzn distribution (Fig.10) on the interval February 1–March 31, 2010, where the threshold limit of about 1.842 separates also two domains, one with normal trend (earthquakes of M<3.4) extended on the interval February 01- February 18, and anomalous one, on the interval February 21- March 31, having Bzn values between 1.850- 1.855. The last interval could be correlated with the superposition effect produced by the two earthquakes of M4.2, and the pre-seismic lead time is about 10 days before the first earthquake of M4.2 occurred.

Figures 11 and 12 depict results of Bzn distribution observed at GOPS on the two intervals May 28 - August 26, 2009 and the whole September month, 2009.

Figure 11 reveals three anomalous domains of Bzn which may be related to 5 earthquakes with magnitude larger than 4. First domain, extended on the interval June 4 – July 10, is characterized by enhanced values of Bzn comprised between 1.852 and 1.854, and may be related to the superposition effect generated by the two earthquakes of M3.9 (June 20) and M4.0 (June 27). Pre-seismic increased values of Bzn are extended on 16 days interval for this group of earthquakes.

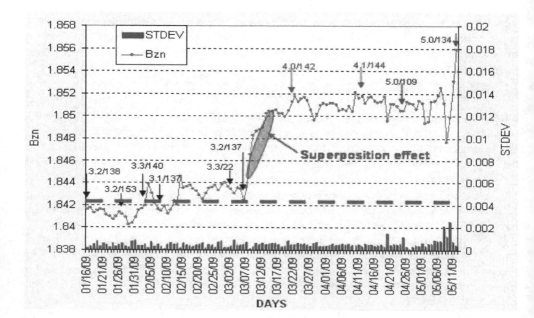

Fig. 9. Bzn and STDEV distributions at the GOPS, within the interval January 16 – April 30, 2009; vertical arrows are earthquakes and ratio 5.0/109 is the magnitude/hypocenter depth of earthquake in [km]; dashed red line is threshold limit between normal trend and anomalous behaviour of the normalized function Bzn

The second anomalous domain, with an average value of Bzn of about 1.855, is extended on the interval July 14 – August 7 and reflects also superposition effect of the two earthquakes of magnitude 5.1 and 5.2, occurred on July 24 and August 5, respectively. The pre-seismic superposition effect of the Bzn started on July 15 and is developed in 9 days interval up to the occurrence of M5.1 earthquake (July 24). In the last anomalous domain (August 11-August 26), the normalized function Bzn has values between 1.853 -1.454 and could be associated with the two earthquakes of M4.1 occurred on August 17 and 26. Here, the pre-seismic occurrence interval is of about 7 days.

Figure 12 illustrates the Bzn distribution for September 2009, where similar pre-seismic characteristics are observed. Thus, enhanced values of Bzn are correlated with the increased values of earthquake magnitudes and decreased foci depth.

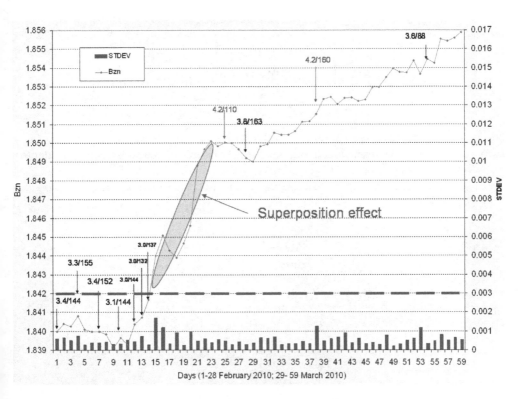

Fig. 10. Bzn and STDEV distributions at the GOPS, within the interval February01– March 31, 2010; vertical arrows are earthquakes and ratio 4.2/110 is the magnitude/hypocenter depth of earthquake in [km]; dashed red line is threshold limit between normal trend and anomalous behaviour of the normalized function Bzn

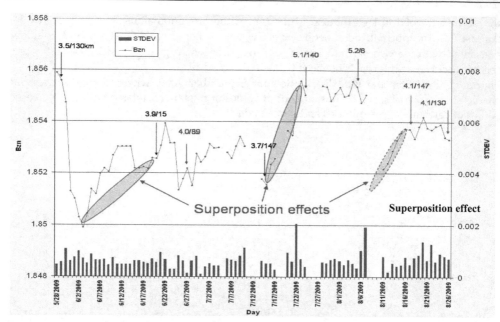

Fig. 11. Bzn and STDEV distributions at the GOPS, within the interval May 28– August 26, 2010; vertical arrows are earthquakes and ratio 5.1/140 is the magnitude/hypocenter depth of earthquake in [km]

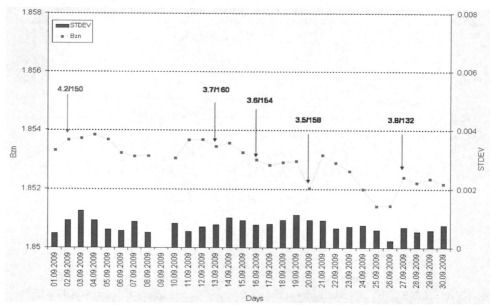

Fig. 12. Bzn and STDEV distributions at the GOPS, within the interval 01February – March 31, 2010; vertical arrows are earthquakes and ratio 4.2/150 is the magnitude/hypocenter depth of earthquake in [km]

The local variation of earthquakes energy (Es) carried out for the analyzed interval in 2009 year is shown on Fig. 12. The relationship between earthquake magnitude and energy in foci is:

$$logEs = 11.8 + 1.5\,M \tag{11}$$

Where: E_S is energy [Erg], M is earthquake magnitude.

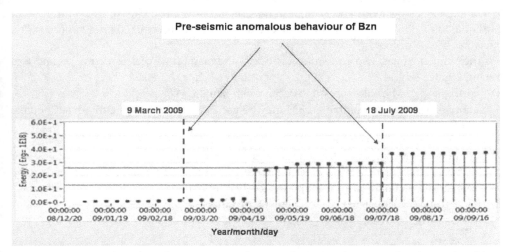

Fig. 13. Variation of earthquakes energy (in foci) on the interval January 01– September 30, 2009: vertical red dashed lines represent lead time of the pre-seismic anomalous behaviour of the normalized function Bzn

It is quite obvious that the pre-seismic anomalous behaviour of Bzn observed on March 9 (Fig.9) and July 18 (Fig.10) respectively, may be correlated with the increased seismic energy reflected by Fig. 13, as follows: more than 15 days interval for the earthquake of M4.0 (March 21) and, about 7 days for earthquakes of M5.1 (July 24). This seismic activity support a possible generation mechanism of the electromagnetic precursors based on the stress generation followed by dehydration of rocks and fluid migration through the faulting system, which may produce concentration of induced currents in highly conductive elongated structure such as Carpathian electrical conductivity anomaly. These induced currents may cause distortion of the vertical geomagnetic component (Fig. 8) which is reflected by increased values of Bzn (Fig.9). The pre-seismic lead time of the normalized function Bzn is between 7days and 15 days for all the data presented in this paper.

5. Conclusions

The results carried out in this paper are based on the hypothesis according to the pre-seismic conductivity changes, due to the fluid migration through faulting system, may generate increased values of the normalized function Bzn proportionally with the intensity of anomalous current concentrations through CECA.

The normalized function Bzn carried out at GOPS has been analyzed in order to detect its pre-seismic anomalous behaviour related to the intermediate depth earthquakes with $M \geq 4$. Before all the earthquakes of $M \geq 4$ the Bzn distribution exhibit significant enhancements from the normal trend and the pre-seismic lead time is about 7-15 days. The Bzn average value of 1.842 (Fig.9), associated with earthquakes of $M \leq 3.3$ occurred on the period January 16 – March 7, represents the threshold limit between the normal trend and its pre-seismic anomalous behaviour, taken as possible earthquake precursor.

When anomalous behavior and normal trend domains are much closed, as a multitude of earthquakes of different magnitude occurred at short time intervals, then a superposition effect has been observed.

For the Vrancea zones, two correlations between the magnitude of seismic events and Bzn are highlighted (Fig.14):

i.   earthquake of $M \geq 4$ is expected to occur when $Bzn \geq 1.847$;

ii. anomalous behaviour of $Bzn \geq 1.854$ may be use as pre-seismic value for an earthquake of $M \geq 5$.

Sometimes, superposition effects may generate exceptions to the above rules.

As this methodology allows us to know always the structure changes after any seismic event (on the base of dimensionality parameters), what permit to use further on the most adequate electromagnetic techniques, it becomes an interesting subject of studying the earthquake mechanism and the associated electromagnetic precursors.

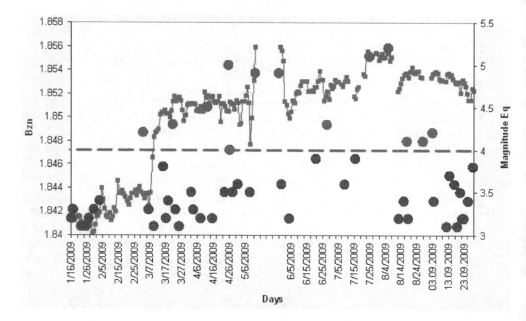

Fig. 14. Distributions of the normalized function Bzn and earthquake magnitude on the 2009 analyzed interval; blue circles are earthquakes with M< 4; red circles are earthquakes with M ≥4; red dashed line represents the limit above to which one earthquake of M ≥4 is possible to occur.

6. Acknowledgment

This work was supported by the CNCSIS - UEFISCDI, project number 1028/2009, PN II – IDEI code 14/2008. We would also like to thank to the Institute of Geodynamics of the Romanian Academy for providing electromagnetic data and facilities for the present research project.

7. References

Bahr, K. (1988). Interpretation of the magnetotelluric impedance tensor: regional induction and local telluric distortion, *J. Geophys.*, 62, pp.119-127

Brace, W.F.; Orange, A. S. & Madden, T.N. (1965). The effect of pressure on the electrical resistivity of water-saturated crystalline rocks. *Journal of Geophysical Res.*, 70 (22), pp. 5669-5678

Fenoglio, M.A.; Johnston M.J. & Bierlee, J.D. (1995). Magnetic and electric fields associated with changes in high pore pressure in fault zones: application to the Loma Prieta ULF emissions. *J.Geophys.Res.* 100, pp. 12951-12958

Fraser-Smith, A.C.; Bernardi, A.; Mc Gill, P.R.; Ladd, M. E.; Halliwell, R.A. & Villard, Jr. O.G. (1990). Low frequency magnetic field measurements near the epicenter of the M 7.1 Lomma Prieta earthquake. *Geophys. Res. Lett.*, 17, pp.1465-1468.

Freund, F.; Gupta, A.; Butow, S.J. & Tenn, S. (1999). Molecular hydrogen and dormant charge carriers in minerals and rocs, In: *Atmospheric and ionospheric electromagnetic phenomena associated with earthquakes*, Hayakawa, & Molchanov, pp. 839-871, Terra Sci. Publ. Comp., Tokyo

Freund, F. (2000). Time-resolved study of charge generation and propagation in igneous rocks. *Journal of Geophysical Res.*, B, 105, pp.11001-11019

Girbacea, R. & Frisch, W. (1998). Slab in the wrong place: Lower lithospheric mantle delamination in the last stage of the Eastern Carpathian subduction retreat. *Geology* 26, pp.611-614

Gotoh, K.; Akinaga, Y.; Hayakawa, M. & Hattori, K .(2002). Principal component analysis of ULF geomagnetic data for Izu island earthquake in July 2000. *J.Atms. Electr.* 22, pp.1-12

Gvirtzman, Z. (2002). Partial detachment of a lithospheric root under the southeast Carpathians: toward a better definition of the detachment concept. *Geology* 30, pp. 51-54

Hattori, K..; Serita, A.; Yoshino, C. Hayakawa, M. & Isezaki, N. (2006). Singular spectral analysis and principal component analysis for signal discrimination of ULF geomagnetic data associated with 2000 Izu Island earthquake swarm. *Phys. Chem. Earth*, 31, pp. 281-291

Hayakawa, M. & Fujinawa, A.Y. (1994). *Electromagnetic Phenomena related to Earthquake Prediction*, Terra Scientific Pub. Comp, Tokyo

Hayakawa, M. & Molchanov, O.A. (2002). *Seismo-Electromagnetics: Lithosphere-Atmosphere-Ionosphere Coupling*, Terra Scientific Pub. Comp., ISBN 9784887041301, Tokyo

Ismail-Zadeh, A.T.; Panza, G.F. & Naimark, B.M. (2000). Stress in the descending relic slab beneath the Vrancea region, Romania, *Pure Appl. Geophys.*, 157, pp.111-130

Johnston M.J.S. (1978). Local magnetic field variations and stress changes near a slip discontinuity on the San Andreas fault. *Journal of Geomagnet. And Geoelec.*, 30, pp. 511-548.

Johnston, M.J.S. (1997). Review of electric and magnetic fields accompanying seismic and volcanic activity. *Surv. Geophys.*, 18, pp. 441-475

Jones, F. W. & Price, A.T. (1970). The perturbations of alternating geomagnetic fields by conductivity anomaly. *Geophys.J.R. Astr. Soc.*, 20, pp. 317-334

Kaufman, A. A. & Keller, G.V. (1981). *The magnetotelluric sounding method,* Elsevier Scientific Publishing Comp., Amsterdam-Oxford-New York.

Khain, V.E. & Lobkosky, L.I. (1994). Conditions of Existence of the Residual Mantle Seismicity of the Alpine Belt in Eurasia, *Geotectonics*, 2, pp. 54-60.

Kopytenko, Y.A.; Matiashvili, T.G.; Voronov, P.M. & Kopytenko, E.A. (1994). Observation of electromagnetic ultralow-frequency lithospheric emission in the Caucasian seismically active zone and their connection with earthquakes, In: *Electromagnetic Phenomena related to Earthquake Prediction*, Hayakawa & Fujinawa, 175-180, Terra Scientific Pub. Comp, Tokyo

Linzer, H.-G. (1996), Kinematics of retreating subduction along the Carpathian arc, Romania, *Geology* 24 (2), pp. 167-170

Liu, J.Y.; Chuo, Y.J.; Shan, S.J.; Tsai, Y.B.; Chen Y.I.; Pulinets, S.A. & Yu, S.B. (2004). Pre-earthquake ionospheric anomalies registered by continuous GPS TEC measurements. *Annals Geophysicae*, 22, pp.1585-1593

Molchanov, O.; Kulchitschy, A. & Hayakawa, M. (2001). Inductive seismo-electromagnetic effect in relation to seismogenic ULF emission. *Natural Hazards and Earth System Sciences*, 1, pp. 61-68

Oncescu, M.C. (1984). Deep structure of the Vrancea region, Romania, inferred from simultaneous inversion for hypocenters and 3-D velocity structure. *Ann. Geophys.*, 2, pp. 23-28

Oncescu, M.C.; Burlacu, V.; Anghel, M. & Smalberger, V. (1984). Three-dimensional P-wave velocity image under Carpathian Arc. *Tectonophysics*, 106, pp. 305-319.

Ouzounov, D.; Bryant, N.; Logan, T.; Pulinets, S. & Taylor, P. (2006). Satellite thermal IR phenomena associated with some of the major earthquakes in 1999-2003. *Physics and Chemistry of the Earth*, 31, pp.154-163

Parck, S.K.; Johnston, M.J.S.; Madden, T.R.; Morgan, F.D. & Morrison, H.F. (1993). Electromagnetic precursors to earthquakes in the ULF band: a review of observations and mechanisms. *Rev.Geophys.*, 31, 117-132.

Parkinson, W. (1959). Directions of rapid geomagnetic variations. *Geophys. J. R. Astr. Soc.* 2, pp.1-14

Parrot, M.; Manninen, J.; Santolik, O.; Nemec, F.; Turner, T.; Rait, T. & Macusova, E. (2007). Simultaneous observations on board of satellite and on the ground of large-scale magnetospheric line radiation. *Geophys. Res. Lett.* 34. L19102

Pinna, E.; Soare, A.; Stanica, D. & Stanica, M. (1993). Carpathian conductivity anomaly and its relation to deep substratum structure. *Acta Geodaet. Geophys. et Montanistica*, Vol. 27(1), pp.35-45

Pulinets, S.A.; Gaivaronska, T.A.; Leyva-Contreras, A. & Ciraolo, L. (2004). Correlation analysis technique reveling ionospheric precursors opf earthquakes. *Nat. Hazards Syst. Sci.*, 4, pp. 697-702

Revol, J.; Day, R. & Fuller, M. (1977). Magnetic behavior of magnetite and rocks stressed to failure-relation to earthquake prediction. *Earth Planet.Sci.Lett.*, 37, pp.296-306.

Rokityansky, I.I. & Ingerov, A.I. (1999). Conductive structure of Ukrainian Carpathians from EM observations. *Phys. Chem. Earth (A)*, 24 (9), pp. 849-852

Sperner, B.; Lorenz, F.; Bonjer, K.; Hettel, S.; Muller, B. & Wenzel, F. (2001). Slab break-off-abrupt cut or gradual detachment? New insights from the Vrancea region (SE Carpathians, Romania). *Terra Nova*, 13, pp.172-179

Sperner, B., the Collaborative Research Centre [CRC], 461 Team (2005). Monitoring of slab detachment in the Carpathians, In:*Challenges for Earth Sciences in the 21st Century*, Wenzel, pp.187-202, Springer-Verlag, Heidelberg

Stanica, D.; Stanica, M.; Piccardi, L.; Tondi, E., & Cello, G. (2004). Evidence of geodynamic torsion in the Vrancea zone (Eastern Carpahians). *Rev. Roum. GEPHYSIQUE*, Vol. 48, pp.15-19

Stanica, D., & Stanica, M. (2007). Electromagnetic monitoring in geodynamic active areas. *Acta Geodinamica et Geomaterialia*, Vol.4, No.1(145), pp. 99-107

Stanica, D., & Stanica, D.A. (2010). Constraints on Correlation Between the Anomalous Behaviour of Electromagnetic Normalized Functions (ENF) and the Intermediate Depth Seismic Events Occurred in Vrancea Zone (Romania). *Terr. Atmos.Ocean.Sci.*,Vol.21, pp.675-683

Stanica, M.; Stanica, D., & Marin-Furnica, C. (1999). The placement of the Trans-European Suture Zone on the Romanian Territory. *Earth Planets Space*, 51, 1073-1078

Tramutoli, V.; Cuomo, V; Fillizzola, C., & Pietrapertosa, C. (2005). Assessing the potential of thermal infrared satellite surveys for monitoring seismically active areas. The case of Kocaely (Izmit) earthquake, August 17-th, 1999. *Remote Sensing of Environment*, 96, pp.409-426

Trifu, C. I., & Radulian, M. (1989). Asperity distribution and percolation as fundamentals of earthquakes cycles. *Phys. Earth Planet. Inter.*, 58, pp.277-288

Tronin, A.A.; Biaggi, P.F., & Molchanov, O.A. (2004). Temperature variations related to earthquakes from simultaneous observations at the ground stations and by satellites in Kamchatka area. *Physics and Chemistry of the Earth*, Vol. 209, pp. 501-506

Wannamaker, P.E.; Stodt, J.A., & Rijo, L. (1986). A stable finite element solution for two-dimensional magnetotelluric modeling. *Geophys. J. R. Astr. Soc.*, 88, pp. 277-296

Varotsos, P. (2005). *The Physics of Seismic Electric Signals* (2005), TERRAPUB, ISBN 4-88704-136-5,Tokyo

Wiese, H. (1962). Geomagnetische tiefensondierung. Teil II: Die Streichrichtung der Undergrundstrukturen des elektrischen Widerstandes, erschlossen aus geomagnetischen variationen. *Geofis. Pura et Appl.*, 52, pp.83-103

Word, R. D.; Smith, H. W., & Bostick Jr., F. X. (1970). *An investigation of the magnetotelluric tensor impedance method*, Electronics Research Center, The University of Texas at Austin, Austin, Texas 78712

Wortel, M.J.R., & Spakman, W. (2000). Subduction and slab detachment in Mediterranean-Carpathian region. *Science,* 290, pp. 1910-1917

5

Identification of Simultaneous Similar Anomalies in Paired Time-Series

R. G. M. Crockett
University of Northampton
United Kingdom

1. Introduction

Changes in radon and other soil-gas concentrations, and other parameters, before and after earthquakes have been widely reported (Asada, 1982; Chyi et al., 2001; Climent et al., 1999; Crockett et al., 2006a; Crockett & Gillmore, 2010; Igarishi et al., 1995; Kerr, 2009; Koch & Heinicke, 1994; Planinic et al., 2000; Plastino et al., 2002; Wakita, 1996; Walia et al., 2005; Walia et al., 2006; Zmazek et al., 2000). However, in the majority of such radon cases, changes in magnitude in single time-series have been reported, often large changes recorded using integrating detectors, and the majority of radon time-series analysis is reported for single time-series (e.g. Baykut et al., 2010; Bella & Plastino, 1999; Finkelstein et al., 1998). With a single time-series, recorded at a single location, there is no measure of the spatial extent of any anomaly and, to a great extent, only anomalies in magnitude can be investigated. With two, or more, time-series from different locations, it is possible to investigate the spatial extent of anomalies and also investigate anomalies in time, i.e. frequency and phase components, as well as anomalies in magnitude.

The aim of this chapter is to present techniques, developed and adapted from techniques more familiar in the field of signal analysis, for investigating paired time-series for simultaneous similar anomalous features. A paired radon time-series dataset is used to illuminate these techniques. This is not to imply that the techniques are restricted to radon time-series: it is simply that the investigation at the University of Northampton of these techniques in the context of earthquake precursory phenomena has been conducted on radon datasets. This work commenced in the autumn of 2002, following the Dudley earthquake of 23 September which was felt in Northampton and which occurred approximately three months into a radon monitoring programme being conducted as part of another project (Crockett et al., 2006a; Phillips et al., 2004).

1.1 UK earthquakes

The UK is not generally regarded as a seismically active region. In general, across the UK as a whole, in any given year there might be a few earthquakes of magnitude up to 3 or 4 and every 5-10 years there might be an earthquake of magnitude 5 or thereabouts (e.g. Bolt, 2004; Musson, 1996). This is simultaneously an advantage and disadvantage to this research. It is an advantage in that with so few earthquakes there is very little seismic 'noise' in any radon, or other, dataset. It is a disadvantage in that with so few earthquakes, long intervals can elapse between events and there is an element of luck in obtaining suitably paired time-series to investigate for potential earthquake-related anomalies.

Indeed, in 2002, luck played a major role in stimulating this research. During the latter part of 2002 the University of Northampton Radon Research Group had two hourly-sampling radon detectors deployed for a period which included the Dudley earthquake and also an unusual earthquake swarm in the Manchester area. Subsequently, 5.5 years then elapsed until another UK earthquake of similar magnitude occurred at Market Rasen in February 2008. Again, the University of Northampton had two hourly-sampling radon detectors deployed and, although that paired time-series was shorter in duration, it was still possible identify simultaneous similar anomalies (Crockett and Gillmore, 2010).

1.2 Magnitude anomalies, probability of occurrence

An anomaly in magnitude is, expressed straightforwardly, a magnitude that occurs infrequently, at low probability, often determined according to user-defined probability criteria in a given context. One commonly used criterion assumes that the magnitudes are normally distributed and defines an anomalous magnitude as being one that lies more than a specified number of standard deviations from the arithmetic mean. For example, in any normally-distributed data, an interval of two standard deviations either side of the mean includes 95.45% of the data and any data lying outside this interval can be defined as anomalous, occurring only 4.55% of the time (2.28% at each tail of the distribution). This straightforward type of criterion is clearly satisfactory for normally distributed data but becomes increasingly less robust with divergence of data from normal distributions, indicating the use of more rigorous probability criteria.

1.2.1 Standardised data, Standardised Radon Index (SRI)

Where the data are not normally distributed, or not sufficiently close to normally distributed in a given context, account must be taken of the specific probability distribution. In some cases, it is possible to map onto a normal distribution, via an equiprobability mapping, and then use the mapped-onto distribution to investigate magnitude anomalies. This is essentially the approach taken with, for example, Standardised Precipitation Indices (SPIs) as described by McKee et al. (1993) which are a representation of (precipitation) data in terms of standard normal variables, i.e. standard deviations around a zero mean. Alternatively, where the data are lognormally distributed, as is generally the case with radon datasets, or for example, square-root or cube-root normally distributed (Fu et al., 2010), it is possible to define magnitude anomalies in terms of the normal distributions of the logarithms, square-roots or cube-roots of the data respectively, again a representation of the data in terms of standard normal variables. This is the premise underpinning the Standardised Radon Indices (SRIs) proposed by Crockett & Holt (2011). SRIs are determined from the normally-distributed logarithms of lognormally distributed radon data, a representation of the data in terms of standard normal variables. Thus, a given magnitude of SRI is determined by the probability of occurrence of a given magnitude of radon concentration within a dataset.

In addition to transforming radon data, or other data by extension, such that the familiar normal-distribution definitions of magnitude anomaly can be used reliably, SRIs also allow radon time-series to be compared more fully than by considering relative magnitude as obtained by normalising the data, e.g. scaling the data to unit mean value. Such normalisation does not account for different radon responses to identical stimuli as might arise from differences in emission properties of rocks, soils, groundwater etc. SRIs, in standardising according to probability of occurrence, allow comparison of different radon datasets having different, possibly non-linear, responses to changes in radon emission in response to identical stimuli. In the case of paired time-series, as being considered herein, if values in the time-series have the same probabilities of occurrence they will have the same SRI even if their (relative) magnitudes are different.

Whilst not the focus of this chapter, this technique is discussed briefly, with an example, in Section 3.

1.3 Quality of data, validity of comparison

In any application of any of the correlation and coherence techniques described herein, the durations and sampling intervals of the paired time-series must be equal. Ideally, the two time-series should be sampled at the same times so as to avoid a built-in time-difference between pairs of data. Where there is such a time difference, it might be possible to pre-process the time series, e.g. via a moving average, to minimise its significance, although this must be balanced against the resultant loss of high-frequency content. Under some circumstances where the sampling intervals are different but one is a multiple of the other, it can be possible to aggregate shorter-interval data to correspond to longer-interval data.

In the following sections, it is implicitly assumed that differences between the paired time-series do not arise from different monitoring equipment responses. Where both time-series are of the same parameter or process recorded using the same equipment, this is generally a safe assumption. However, where this is not the case, e.g. the same parameter or process recorded using different equipment for each time-series or different parameters or processes (recorded using different equipment), it will generally be necessary to pre-process the time-series to minimise such differences. In such cases, filtering or spectral decomposition techniques can be used. In particular, Empirical Mode Decomposition (EMD), in decomposing a time-series into separate components (Intrinsic Mode Functions) according to frequency content, has shown promise (Crockett & Gillmore, 2010; Feng, 2011; Huang *et al.*, 1998; Rilling *et al.* 2003).

2. Correlation and coherence

Correlation and coherence are techniques for comparing time-series (more generally, waveforms, signals). In brief, correlation compares shape, i.e. envelope, and is a time-domain technique; coherence compares composition, i.e. frequency (harmonic) content, and is a frequency-domain technique. Neither technique directly compares scale, i.e. neither directly detects magnitude anomalies.

Correlation is a relatively familiar and straightforward technique, widely used in various forms to compare datasets in general. However, correlation can be misleading, particularly if used in isolation as sole means of comparison. For example, consider a pair of identical time-series, such as two equal-frequency sinusoids in the simplest case. If the two sinusoids are exactly in-phase, then their correlation coefficient will be 1 (maximum positive correlation) because both time-series are always changing in the same sense (positively or negatively) at the same time. Conversely, if they are exactly, i.e. a half-cycle, out-of-phase then their correlation coefficient will be –1 (maximum negative correlation) because one time-series is always changing in the opposite sense to the other. Depending on their phase difference, their correlation coefficient will be between or equal to these two limiting values. However, if they are a quarter-cycle out of phase, their correlation coefficient will be zero. This is because in consecutive quarter-periods the time-series are alternately changing in the same or opposite sense, yielding alternate quarter-periods of positive and negative correlation, equal in magnitude, which sum to zero over a complete period. Thus, whilst a zero, or small, correlation coefficient can indicate a real lack of similarity between two time-series, it can be misinterpreted if there is no information with regard to their frequency content (and phase relationship).

This leads to the less familiar coherence technique. Coherence measures common frequency content, with or without phase information depending on the exact definition. Strictly, the coherence coefficient (i.e. "coherence") measures similarity of frequency content only, irrespective of relative phase, but the overall coherence analysis readily yields a phase-difference for each frequency considered for the coherence coefficient. In the hypothetical case of paired equal-frequency sinusoids considered above, the coherence would be maximal, i.e. coherence coefficient of 1, in all cases irrespective of the phase difference, although the coherence-derived phase-difference would vary exactly as the actual phase-difference. If there is no common frequency content, the coherence is zero.

Correlation is a time-domain technique, coherence is a frequency-domain technique: they are related via their Fourier transforms. Before considering correlation and coherence, therefore, a brief review of Fourier transforms – discrete Fourier transforms – will be given.

2.1 Fourier transform and Discrete Fourier Transform (DFT)

For a continuous function of time, $x(t)$, of infinite duration its continuous Fourier transform is:

$$X(f) = \int_{-\infty}^{\infty} x(t)e^{-2\pi i f t} dt ,$$ (1)

and it is a frequency-domain representation of it (its spectrum). The inverse transform is:

$$x(t) = \int_{-\infty}^{\infty} X(f)e^{2\pi i f t} df .$$ (2)

In practice, time-series are neither continuous nor infinite, and so the Discrete Fourier Transform (DFT) is used, generally via a Fast Fourier Transform (FFT) algorithm. Subject to constraints arising from the finite and discrete nature of the time-domain function $x(t)$, its DFT, $X(f)$, is a frequency-domain representation of it comprising a spectrum of discrete frequencies on a finite frequency interval. Thus, the forward DFT, $x(t) \rightarrow X(f)$, is:

$$X_n = \frac{1}{N} \sum_{k=0}^{N-1} x_k e^{-\left(\frac{2\pi i}{N}\right)nk} , \text{ for } x(t) = x_0, x_1, ..., x_{N-1}; x_k = x(t_k) ; N \text{ samples} ,$$ (3)

The inverse DFT (IDFT), $X(f) \rightarrow x(t)$, is:

$$x_k = \sum_{n=0}^{N-1} X_n e^{\left(\frac{2\pi i}{N}\right)nk} \text{ , for } X(f) = X_0, X_1, ..., X_{N-1}; \ X_n = X(f_n) \text{ ; } N \text{ elements .} \tag{4}$$

For further information on and fuller descriptions of Fourier transforms and their properties see, for example, Riley, 1974; Gabel & Roberts, 1986; Proakis & Manolakis, 2006.

2.2 Correlation analysis

Covariance and correlation are measures of similarity of shape, correlation is normalised covariance and so is independent of scale (e.g. Cramer, 1946; Mood *et al.* 1974). Generalising from the simple case outlined above, the correlation coefficient between two datasets (or samples) will be +1 or –1 if they have the same shape, in-phase or out-of-phase/sign-inverted respectively. The magnitude of the correlation coefficient will reduce according to differences in shape and, as noted above, mismatch in phase.

More formally, in summary, for the comparison of two N-sample time-series, $\{x_i\}, \{y_i\}$, e.g. to ascertain if they contain similar features possibly occurring with a lag between them, there is lagged covariance, $\sigma_{xy}(k)$, and lagged cross-correlation, $R_{xy}(k)$, i.e.:

$$\sigma_{xy}(k) = \sum_{n=1}^{N} (x_n - \mu_x)(y_{n-k} - \mu_y) \text{ , for } k \le N \text{ ,} \tag{5}$$

$$R_{xy}(k) = \frac{\sigma_{xy}(k)}{\sigma_x \sigma_y}. \tag{6}$$

The straightforward unlagged covariance and (cross-) correlation of two time-series are (5) and (6) evaluated at zero-lag, i.e.:

$$\sigma_{xy} = \sigma_{xy}(0) = \sum_{n=1}^{N} (x_n - \mu_x)(y_n - \mu_y), \tag{7}$$

$$R_{xy} = R_{xy}(0) = \frac{\sigma_{xy}(0)}{\sigma_x \sigma_y}. \tag{8}$$

Where, in equations (5) – (8):

x_i, y_i are the ith members of time-series $\{x_i\}, \{y_i\}$ respectively

μ_x, μ_y are the mean values of $\{x_i\}, \{y_i\}$ respectively

σ_x, σ_y are the standard deviations of $\{x_i\}, \{y_i\}$ respectively

For the comparison of a time-series against lagged versions of itself, e.g. to find the period(s) of any cyclic features, there is (lagged) autocovariance, $\sigma_{xx}(k)$, and (lagged) autocorrelation $R_{xx}(k)$, i.e.:

$$\sigma_{xx}(k) = \sum_{n=1}^{N} (x_n - \mu_x)(x_{n-k} - \mu_x), \tag{9}$$

$$R_{xx}(k) = \frac{\sigma_{xx}(k)}{\sigma_{xx}(0)} = \frac{\sigma_{xx}(k)}{\sigma_{xx}}, \tag{10}$$

which are (5) and (6) with both time-series the same, and clearly both have their maximum values (i.e. autocorrelation is +1) at zero lag.

2.2.1 Rolling/sliding correlation

Whilst lagged cross-correlation reveals any lag between two time-series and cross-correlation of whole time-series gives an overall measure of similarity, neither reveals any time-dependence of similarity during the time-series, i.e. sections where the time-series correlate to greater or lesser extents. One means of achieving this is to window the paired time-series, starting at the beginning, cross-correlate across the window, roll/slide the window forwards a number of samples and repeat the cross-correlation, repeating the roll/slide-and-correlate procedure until the end of the time-series is reached. This yields a time-series of correlation coefficients which reveals those sections of the paired time-series which are varying exactly in phase ($R = 1$), those which are varying exactly out-of-phase ($R = -1$) and all intermediate values. This can be repeated for different window-durations and, for example, the results presented as a contour plot to reveal the time-duration relationships of any periods of significant cross-correlation between the time-series, analogous to the more familiar spectrogram representation of time-frequency relationships in single time-series, and to cross-coherence, as described below (e.g. Crockett *et al.*, 2006a).

2.3 Coherence (cross-spectral) analysis

Coherence (cross-coherence, magnitude-squared coherence) can be useful in that it measures the similarity of two signals, i.e. time-series in this context, in terms of their frequency composition (Brockwell & Davis, 2009; Penny, 2009; Proakis & Manolakis, 2006; Venables & Ripley, 2002). It is a normalised measure of power cross-spectral density and is a frequency-domain measure of correlation of the two signals (time-series).

2.3.1 Power spectral density

The concept of "signal power" is not itself useful in this context but the terminology is established and is retained herein for reasons of simplicity. In signals which transmit energy, power, it is significant and is determined by the square of the amplitude.

The power spectral density is obtained via the Discrete Fourier Transform and is the proportion of the total power content, i.e. square-of-magnitude, carried at given frequencies. As defined by the Wiener-Khintchine Theorem (Proakis & Manolakis, 2006), the power spectral density, G_{xx}, of a signal (time-series) is the Fourier transform of the autocovariance, i.e.:

$$G_{xx}(k) = \sum_{n=0}^{N-1} \sigma_{xx}(n) e^{-i\frac{2\pi}{N}nk} . \tag{11}$$

Also, the power cross-spectral density, G_{xy}, of two signals (time-series) is the Fourier transform of their cross-covariance, i.e.:

$$G_{xy}(k) = \sum_{n=0}^{N-1} \sigma_{xy}(n) e^{-i\frac{2\pi}{N}nk} . \tag{12}$$

2.3.2 Coherence (magnitude-squared coherence)

The coherence (coherence-coefficient), C_{xy}, is the cross-spectral density (complex) normalised by the product of the individual spectral densities (real), i.e.:

$$C_{xy}(k) = \frac{\left|G_{xy}(k)\right|^2}{G_{xx}(k)G_{yy}(k)}, \tag{13}$$

and the phase difference, Φ_{xy}, is given by:

$$\tan(\Phi_{xy}(k)) = \frac{\text{Im}(G_{xy}(k))}{\text{Re}(G_{xy}(k))}. \tag{14}$$

The coherence, C_{xy}, is real and varies between zero, i.e. no common components of the signals (time-series) at frequency k, and 1, i.e. equal proportional components of the signals (time-series) at frequency k.

2.4 Summary

This section has outlined the mathematics of correlation and coherence and also summarised their strengths and weaknesses. Both techniques have their strengths and weaknesses but complement each other. Correlation compares shape (envelope), effectively in-phase periodic features, but gives no information regarding frequency content and so can be misleading with regard to out-of-phase features. Conversely, coherence compares periodic structure (frequency composition and phase difference) and so informs with regard to similarity of composition and structure, accounting for phase difference, but not with regard to overall shape. Thus, used in combination, a much more complete comparison can be obtained than by using either in isolation.

3. Case-study: July – November 2002

The techniques outlined above have been investigated, developed and adapted using primarily one paired time-series dataset: this is a radon dataset comprising two hourly-sampled time-series spanning 5.5 months from late June to mid December 2002. This period also included the M_L=5 Dudley (UK) earthquake of 23 September (22 September GMT), which was widely felt by people in Northampton (and elsewhere in the English Midlands), and the Manchester (UK) earthquake swarm of 21-29 October, which wasn't felt in Northampton but was widely felt in southern parts of NW England and northern parts of the English Midlands. Such events are unusual for the UK and, the Dudley earthquake in particular, were the stimulus for the original investigation (Crockett et al., 2006a).

In effect, this is an investigation of three time-series, one of earthquake incidence and two of hourly-sampled radon concentrations, for common radon responses to earthquake stimuli. The hypothesis for this investigation is that a big disturbance, such as an earthquake, occurring at a relatively large distance compared to the detector separation could be expected to produce simultaneous similar radon anomalies (Crockett et al., 2006a). Conversely, any anomalies which are neither simultaneous nor similar are more likely to arise from stimuli local to individual monitoring locations. This hypothesis is appropriate under the circumstances for the radon time-series considered herein, as described below.

However, under different circumstances, e.g. where the detector separation is greater in comparison to the distance from the stimulus, or where one detector location is significantly closer to the stimulus than the other, the analysis might have to modified appropriately to account for a time lag, e.g. moving one time-series relative to the other in the time domain as indicated by lagged cross-correlation. Also, where the individual time-series arise from different monitored substances or processes (e.g. radon and another soil-gas or rock property), pre-processing as indicated in section 1.3 (or otherwise) might be appropriate.

3.1 The time-series

The radon data were collected using Durridge RAD7s, operated 2.25 km apart in Northampton (Phillips *et al.*, 2004; Crockett *et al.*, 2006a,b). A central 20-week extract, 14 July – 30 November 2002, of the 5.5-month paired time-series is shown in Figure 1.

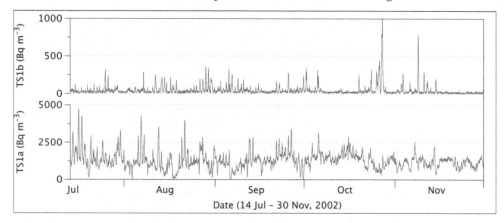

Fig. 1. The paired radon time-series, central 20-week period.

In summary, both time-series are lognormally distributed, with correlation coefficients to lognormal distributions of > 0.91, although TS1a recorded radon levels typically 30-40 times greater than TS1b (Crockett *et al.*, 2006a). Both time-series are characterised by weak, noisy, non-stationary 24 h cycles having short autocorrelation times, *ca.* 1-2 days, as shown in the spectrograms and autocorrelograms in Figures 2 and 3 respectively. There is also weak evidence of 7 day cycles, typical of anthropogenic influences, and in TS1b there is evidence of some longer-period variations, durations *ca.* 15 and 30 days, which are possibly attributable to lunar-tidal influences (Crockett *et al.*, 2006a, 2006b). With the exception of rainfall, behind which TS1a and TS1b lag by 14 and 10 days respectively, there are no observed meteorological dependencies (Crockett *et al.*, 2006a).

The spectrograms are for 6 day windows, stepped at 1 day intervals. This window duration gives the best balance between resolution in the time and frequency domains in light of the durations of the periods of high correlation and coherence discussed below.

The earthquake data for the monitoring period, for earthquakes occurring within 250 km of Northampton, are presented in Table 1 (BGS (http://www.bgs.ac.uk) 2003; USGS/ANSS (http://quake.geo.berkeley.edu/anss), 2011). The next nearest earthquakes during this period occurred at distances greater than 420 km, i.e. there is a distance interval of 170 km between the earthquakes listed and the next nearest earthquakes. As well as the Dudley and Manchester earthquakes, there were other earthquakes of interest, these being an English

Channel earthquake on 26 August and a North Sea earthquake on 22 November (not of identified interest in previous stages of this research).

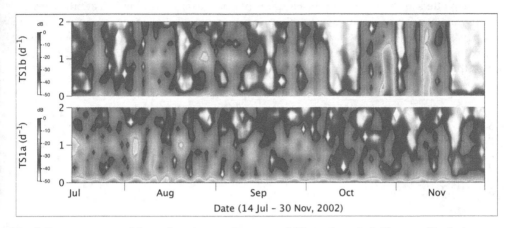

Fig. 2. Spectrograms of the radon time-series, central 20-week period. The amplitude is calibrated in dB, from maximum 0 dB (red) to –50 dB (blue-white). The vertical axes are cycles per day.

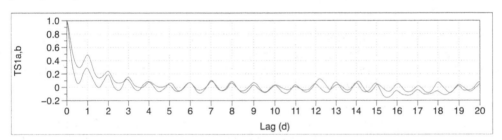

Fig. 3. Autocorrelograms of the radon time-series, central 20-week period (TS1a in red, TS1b in blue).

Date / Time (GMT)	Lat.	Lon.	Depth (km)	Mag. (M_L)	Dist. (km)	Location
26/08/2002 23:41	50.048	-0.009	4.0	3.0	247	Eng. Channel
22/09/2002 23:53	52.520	-2.150	9.4	5.0	94	Dudley
23/09/2002 03:32	52.522	-2.136	9.3	3.2	93	Dudley
21/10/2002 07:45	53.475	-2.000	5.0	3.7	161	Manchester
21/10/2002 11:42	53.478	-2.219	5.0	4.3	169	Manchester
22/10/2002 12:28	53.473	-2.146	4.2	3.5	165	Manchester
23/10/2002 01:53	53.477	-2.157	5.0	3.3	166	Manchester
24/10/2002 08:24	53.485	-2.179	3.7	3.8	168	Manchester
29/10/2002 04:42	53.481	-2.198	5.0	3.1	168	Manchester
22/11/2002 01:40	52.921	2.430	10.0	3.4	237	North Sea

Table 1. Earthquakes ($M_L \geq 3$) within 250km of Northampton, July – December 2002.

3.2 Correlation results

The rolling/sliding windowed correlation, for windows of duration 1-10 days, is shown contour-plotted in the upper plot of Figure 4 (vertical axis is window duration), with the radon time-series and earthquake timings also shown in the lower plot. In this and subsequent figures the time-series are shown normalised to unit mean, to assist visual comparison. This is a simple scaling of amplitudes: in terms of the waveforms of the time-series, the shape and phase are preserved unaltered by this normalisation.

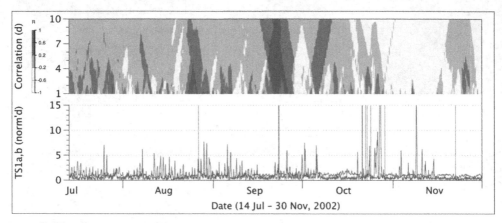

Fig. 4. Rolling/sliding windowed cross-correlation shown contour-plotted (upper plot, strong red and pale yellow areas indicate high positive and negative correlation respectively) with time-series and earthquake incidence (lower plot, TS1a and TS1b in red and blue respectively, earthquake timings as vertical black lines).

Figure 4 (upper) shows two distinct periods of high positive correlation (i.e. red); around (i) 21-23 September, across window durations of up to 10 days, and (ii) 25-27 August, across window durations of up to 5-6 days. The correlation coefficient for the whole period is $R = -0.08$ and the mean values of the correlation coefficient across all ten window durations is $|R| \leq 0.036$. Therefore, the paired time-series typically do not correlate and periods of high correlation are anomalous: e.g. the maximum values of the correlation coefficient for 1-5 day windows occur less than 5% of the time and do so predominantly in these two periods. The more significant of these two periods, 21-23 September, corresponds temporally to the Dudley earthquake of 22 September. The other period, 25-27 August, corresponds temporally to the English Channel earthquake of 26 August. Each period lasts $ca.$ 5-7 days and, in each, correlation peaks prior to the earthquakes by $ca.$ 1 day. There is no similar period of high positive correlation at the time of the Manchester or North Sea earthquakes.

3.3 Coherence results

The coherence, for windows of duration 6 days stepped at 1 day intervals, is shown in Figure 5. The corresponding phase coherence is shown in Figure 6 (phase difference of TS1b relative to TS1a). In these (upper) figures, the coherence information is contour-plotted to show the variation in coherence (Fig. 5) and phase-difference (Fig. 6) at frequencies of 0-3 cycles per day (vertical axes). The 6 day window duration corresponds to the window duration used for the spectrograms and also to the durations of the two periods of high positive correlation.

In both coherence and phase-coherence, it is the 24 h cycles, as revealed by the spectrograms and autocorrelograms, which are the focus as these constitute the main frequency component in each time series and, thus, the main frequency at which the time-series could cohere. The figures also show coherence and phase-coherence for 12 h cycles, i.e. twice the frequency of the main cycles, because coherence at this frequency also reveals some information. There is no significant coherence information outside this frequency range.

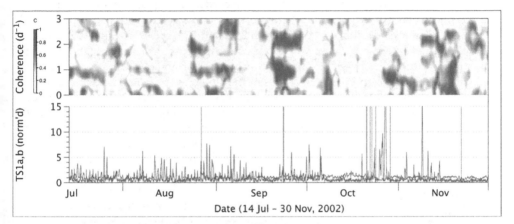

Fig. 5. Coherence shown contour-plotted (upper plot, strong red and pale yellow areas indicate high and low coherence respectively) with time-series and earthquake incidence (lower plot, TS1a and TS1b in red and blue respectively, earthquake timings as vertical black lines).

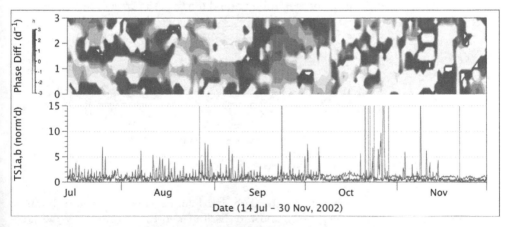

Fig. 6. Phase coherence shown contour-plotted (upper plot, strong red and pale yellow areas indicate positive and negative phase difference respectively, mid-orange indicates small phase difference) with time-series and earthquake incidence (lower plot, TS1a and TS1b in red and blue respectively, earthquake timings as vertical black lines).

Figure 5 (upper) shows two conspicuous periods of high coherence (i.e. red) for both 24 h and 12 h cycles; i.e. late August, late September, and another less conspicuous period in mid-late November. The late August and late September periods correspond to the periods of high positive correlation shown in Figure 4 and correspond temporally to the English Channel and Dudley earthquakes respectively. The mid-late November period occurs a few days before the North Sea earthquake of 22 November. Additionally, there is a period of high coherence for 24 h cycles in late October, which corresponds temporally to the end of the Manchester earthquake swarm, and also a period of high coherence for 24 h cycles in late July with no apparent correspondence to any recorded earthquakes in the region.

Figure 6 (upper) shows two periods where the time-series are in phase (i.e. approximately zero phase difference, medium orange) for both 24 h and 12 h cycles, one in late August and the other in late September. These two periods confirm the two periods of high correlation: if the principal periodic content (24 h) and a principal harmonic (12 h) are in phase then there will be underlying in-phase similarities in the envelopes, giving rise to (high) positive correlation. However, across the time-series as a whole, the daily maxima typically occur at *ca.* 18:00 and 16:00 GMT for TS1a and TS1b respectively, i.e. typically the time-series are not in-phase. This is confirmed directly in Figure 6, which shows variations in phase-difference throughout the period, and also confirmed by the short autocorrelation times shown in Figure 3 and the small average cross-correlation coefficients shown in Figure 4.

3.4 Summary of results
The rolling/sliding windowed correlation, as reported in the initial investigation (Crockett *et al.* 2006a), reveals strong evidence that the two radon time-series correlate positively at the time of the Dudley earthquake, and also evidence that they correlate positively at the time of the English Channel earthquake, but that in general they do not correlate. These results are confirmed by the coherence results, both coherence coefficients and phase-difference.

Seven day details from the time-series are shown in Figure 7, for the period around the Dudley earthquake and Figure 8, for the period around the English Channel earthquake.

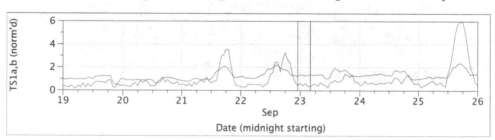

Fig. 7. Late September Anomalies (TS1a in red, TS1b in blue, earthquake timings as vertical black lines).

The two periods are similar in showing the daily maxima, i.e. spikes of *ca.* 6 h duration, occurring simultaneously (i.e. in-phase), and are dissimilar to the remainder of the time-series. Some of these simultaneous maxima occur before the earthquakes – and are potentially earthquake precursory phenomena.

It is clear from these figures, however, that the relative magnitudes of these daily maxima are 2-4 times greater in TS1b than TS1a, indicating that the radon emission characteristics in the two locations are different. The reasons for this will be differences in the rocks and soils at the two locations and possibly different ventilation characteristics in the two basements where the RAD7s were placed. However, if the time-series are represented in terms of SRIs, effectively standard normal variables, then these different responses are (partially) equalised in terms of probability of occurrence. This is shown in Figure 9 for the September anomalies.

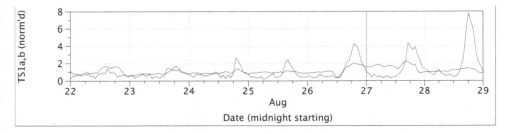

Fig. 8. Late August Anomalies (TS1a in red, TS1b in blue, earthquake timing as a vertical black line).

Figure 9 shows that these simultaneous daily maxima are more similar in probability than in relative magnitude. In SRIs, these maxima have magnitudes *ca.* 1.5-2.5, i.e. 1.5-2.5 standard deviations from the mean. Thus, some of these would 'pass' a plus-or-minus 2 standard deviations criterion for magnitude anomaly whilst others would not, and none would 'pass' a plus-or-minus 3 standard deviations criterion. Therefore, what the correlation and coherence analyses have identified are anomalies in the time domain, i.e. anomalies in frequency composition and phase, rather than in the magnitude domain.

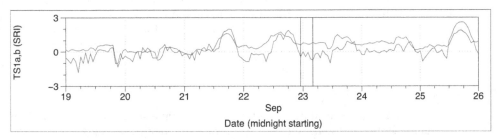

Fig. 9. Late September Anomalies (TS1a in red, TS1b in blue, earthquake timings as vertical black lines).

The late September and late August periods are identified as time-domain anomalies by correlation, coherence and phase-coherence (at 24 h and 12 h cycles), and these periods temporally correspond to the Dudley and English Channel earthquakes. The mid-late November period of coherence (at 24 h and 12 h cycles), but without significant correlation or phase-coherence, is another time-domain anomaly, which has similarities to (coherence) and differences from (correlation, phase-coherence) the late September and late August time-domain anomalies. This mid-late November anomaly occurs a few days prior to the 22 November earthquake but any temporal association with that earthquake is weakened by the absence of corresponding correlation or phase-coherence information. The two periods of 24 h coherence (without 12 h coherence) are clearly ambiguous: the late October period corresponds temporally to the later Manchester earthquakes but the late July period does not correspond temporally to any recorded earthquake in the region.

Noting the relative magnitudes and proximities of the Manchester earthquakes compared to the English Channel earthquake, it is perhaps surprising the there is no similarly well-defined time-domain anomaly in the radon time-series at the time of Manchester earthquakes. The reasons for this are not currently understood but there are essentially two possibilities. First, any such temporal correspondence between earthquake and time-domain anomalies in the radon time-series is coincidental, as discussed below. Second, the temporal correspondence is not coincidental but the natures of the geologies at and between Northampton and Manchester are such that any earthquake-related radon-stimuli are blurred and attenuated. For information on geology see, for example, Boulton, 1992; Hains & Horton, 1969; Poole et al., 1968; Smith et al., 2000 and Toghill, 2003. However, having also identified time-domain radon anomalies which temporally correspond to the Market Rasen earthquake of 27 February 2008 (Crockett & Gillmore, 2010), the second reason is arguably more probable but more data are required, both earthquake and radon, to investigate this more fully.

In these data, another potential geophysical explanation is possible: i.e. lunar-tidal influences, which have been reported for TS1b in terms of cyclic variations in radon concentration (Crockett et al., 2006b). Tidal influences might account for the periods of 24 h coherence which temporally correspond to tidal maxima associated with the full moons of 24 July, 22 August, 21 September and 20 November, but not the absence of any coherence around the 21 October full moon. This potential explanation is further weakened by the absence of (a) consistent coherence around the new-moon maxima and (b) consistent correlation or phase coherence around both sets of maxima. Also, a tidal-maximum explanation does not account for the anomalies in the February 2008 time-series, which temporally correspond to the Market Rasen earthquake but not a tidal maximum. Lastly, despite the apparent similarities in timing, it is unknown whether there is any lunar-tidal influence on the English Channel, Dudley, Manchester and North Sea earthquakes and no such influence has been reported for UK earthquakes in general.

4. Conclusions

Both correlation and coherence show when two, or more, time-series behave similarly in the time domain, according to shape (correlation) or frequency composition (coherence). Thus, these techniques allow the identification of time-domain anomalies, i.e. periods in time when the common behaviour of two, or more, time-series changes from the typical to the anomalous. In the radon data used to illustrate the techniques, the paired time-series typically neither correlate nor cohere but do so anomalously for short periods. In other data, the emphasis might be different, e.g. the time series might typically both correlate and cohere but contain anomalous periods where they do not or, the time-series might typically cohere at some frequencies but contain anomalous periods where the cohering frequencies change.

Correlation does not imply causality, is not proof of causality: at most, correlation might be evidence to support causality. In the dataset analysed above, despite the clear temporal correspondence of the late September and late August time-domain anomalies to earthquakes, and the temporal correspondence of the mid-late November less well-defined time-domain anomaly to another earthquake, this is all that is shown, i.e. temporal correspondence. The analysis does not prove that the anomalies are related to the earthquakes, i.e. does not demonstrate that an earthquake-stimulus radon-response relationship exists, but such analysis does provide necessary evidence towards the demonstration that such a relationship might exist.

With regard to magnitude anomalies, care must be taken to apply criteria used for identifying anomalies correctly dependent upon the probability distribution(s) of the data being investigated. Noting the simplicity and familiarity of the normal distribution, and associated *de facto* standard criteria for determining anomalies, a technique such as the SRI which maps data onto standard normal variables is useful, but this technique is also useful in effectively equalising different, generally non-linear radon-emission characteristics and facilitating comparison in terms of probability of occurrence.

5. Acknowledgements

The author gratefully acknowledges members of the University of Northampton Radon Research Group for the collection of the data and also DEFRA (UK) for funding the research under which the 2002 data were collected (EPG 1/4/72, RW 8/1/64). The author also acknowledges UNESCO / IUGS / IGCP Project 571 for facilitating preparation and dissemination of earlier stages of this and related research.

The data-analysis was performed using the open-source R (http://www.r-project.org)) and Scilab (http://www.scilab.org) software packages, with the EMD and Seewave R libraries.

6. References

Asada, T. (1982). *Earthquake Prediction Techniques: Their Application in Japan*, University of Tokyo Press, Japan.

Baykut, S., Akgul, T. & Seyis, C. (2010). Observation and removal of daily quasi-periodic components in soil radon data, *Rad. Meas.*, Vol.45, No.7, pp. 872-879, doi:10.1016/j.radmeas.2010.04.002.

Bella, F. & Plastino, W. (1999). *Radon time series analysis at LNGS, II*, LNGS Annual Report 1999, Gran Sasso National Laboratory, Italy, pp.199-203.

Bolt, B.A. (2004). *Earthquakes* (5th edition). W.H. Freeman & Co, New York, USA. ISBN 0-7167-5618-8.

Boulton, G.S. (1992). Quaternary, In: *Geology of England and Wales*, Duff, P.McL.D. & Smith, A.J. (Eds.),The Geological Society, London, pp.413-444.

Brockwell, P.J. & Davis, R.A (2009). *Time Series: Theory and Methods*, Springer. ISBN 978-1-4419-0319-8.

Chyi, L., Chou, C-Y., Yang, F. & Chen, C-H. (2001). *Continuous Radon Measurements in Faults and Earthquake Precursor Pattern Recognition*, Western Pacific Earth Sciences, Vol.1, No.2, pp.227-246.

Climent, H., Tokonami, S. & Furukawa, M. (1999). Statistical Analysis Applied to Radon and Natural Events, *Proc. Radon in the Living Environment*, Athens, Greece, April 1999, pp.241-253.

Cramer, H. (1974, repub. 1999). *Mathematical Methods of Statistics*, Princeton University Press, ISBN 978-06-910-0547-8.

Crockett, R.G.M., Gillmore, G.K., Phillips, P.S., Denman, A.R. & Groves-Kirkby, C.J. (2006a). Radon Anomalies Preceding Earthquakes Which Occurred in the UK, in Summer and Autumn 2002. *Science of The Total Environment*, Vol.364, pp. No.1-3, pp.138-148. doi:10.1016/j.scitotenv.2005.08.003.

Crockett, R.G.M., Gillmore, G.K., Phillips, P.S., Denman, A.R. and Groves-Kirkby, C.J. (2006b). Tidal Synchronicity of Built Environment radon levels in the UK. *Geophys. Res. Lett.* Vol.33, No.5, L0538, doi:10.1029/2005GL024950.

Crockett, R.G.M. & Gillmore, G.K. (2010). Spectral-decomposition techniques for the identification of radon anomalies temporally associated with earthquakes occurring in the UK in 2002 and 2008. *Nat. Hazards Earth Syst. Sci.*, 10, 1079–1084, doi:10.5194/nhess-10-1079-2010.

Crockett, R.G.M. & Holt, C.P. (2011). Standardised Radon Index (SRI): a normalisation of radon datasets in terms of standard normal variables. *Nat. Hazards Earth Syst. Sci.*, 11, 1839-1844, doi:10.5194/nhess-11-1839-2011.

Feng, Z. (2011). The seismic signatures of the 2009 Shiaolin landslide in Taiwan, *Nat. Hazards Earth Syst. Sci.*, Vol.11, 1559-1569, doi:10.5194/nhess-11-1559-2011.

Finkelstein, M., Brenner, S., Eppelbaum, L. & Ne'eman, E. (1998). Identification of Anomalous Radon Concentrations Due to Geodynamics Processes by Elimination of Rn Variations Caused by Other Factors. *Geophys. J. Int.*, Vol.133, No.5, pp.407-412, doi:10.1046/j.1365-246X.1998.00502.x.

Fu, G., Viney, N.R. & Charles, S.P. (2010). Evaluation of various root transformations of daily precipitation amounts fitted with a normal distribution for Australia. *Theor. Appl. Climatol.* Vol.99, No.1-2, pp.229–238, doi:10.1007/s00704-009-0137-6.

Gabel, R.A. & Roberts, R.A. (1986). *Signals and Linear Systems* (3rd edition), Wiley. ISBN 978-04-718-2513-5

Hains, B.A. & Horton, A. (1969). *British Regional Geology Central England.* (3rd edition), HMSO.

Huang. N.E., Shen. Z., Long. S.R., Wu, M.L., Shih, H.H., Zheng, Q., Yen, N.C., Tung, C.C., Liu, H.H. (1998). The empirical mode decomposition and Hilbert spectrum for nonlinear and nonstationary time series analysis. *Proc. Roy. Soc. A*, Vol.454, No.1971, pp.903–995, doi:10.1098/rspa.1998.0193.

Igarashi, G., Saeki, S., Takahata, N., Sumikawa, K., Tasaka, S., Sasaki, Y., Takahashi, M. & Sano, Y. (1995). Ground-Water Radon Anomaly Before the Kobe Earthquake in Japan, *Science*, Vol.269, pp.60-61, DOI: 10.1126/science.269.5220.60.

Kerr, R.A. (2009). After the Quake, in Search of the Science – or even good prediction. *Science*, Vol.324, No.5925, p.322, doi:10.1126/science.324.5925.322.

Koch, U. & Heinicke, J. (1994). Radon Behaviour in Mineral Spring Water of Bad Bramburgh (Vogtland, Germany) in the Temporal Vicinity of the 1992 Rörmond Earthquake, the Netherlands, *Geologie en Mijnbouw*, Vol.73, pp.399-406.

McKee, T.B., Doesken, N.J. & Kleist, J. (1993). The relationship of drought frequency and duration to time scales. Preprints, *8th Conference on Applied Climatology*, pp.179–184. January 17–22, Anaheim, California.

Meyer, L.L. (1977). *California Quake*, Sherbourne Press, Nashville, USA.

Mood, A.M., Graybill, F.A. & Boes, D.C. (1974). *Introduction to the Theory of Statistics* (3rd edition), McGraw-Hill. ISBN 978-00-708-5465-9

Musson, R. (1996). British earthquakes and the seismicity of the UK. *Geoscientist*, Vol. No.2, pp. 24-25.

Penny, W.D. (2009). Signal Processing Course, April 2011, Available from http://www.fil.ion.ucl.ac.uk/~wpenny/course/course.html

Planinic, J., Radolic, V. & Culo, D. (2000). Searching for an Earthquake Precursor: temporal variations of radon in soil and water, *Fizika B*, Vol.9, No.2, pp.75-82. ISSN 1330-0008.

Plastino, W., Bella, F., Catalano, P. & Di Giovambattista, R. (2002). Radon groundwater anomalies related to the Umbria-Marche, September 26, 1997, Earthquakes, *Geofisica Internacional*, Vol.41, No.4, pp.369-375. ISSN 0016-7169.

Phillips, P.S., Denman, A.R., Crockett, R.G.M., Groves-Kirkby, C.J. & Gillmore, G.K. (2004). *Comparative analysis of weekly vs. three-monthly radon measurements in dwellings.* DEFRA commissioned research for radioactive substances division. Report DEFRA/RAS/03.006. ISBN 1-900868-44-x.

Poole, E.G., Williams, B.J. & Hains, B.A. (1968). *Geology of the Country around Market Harborough.* Institute of Geological Sciences Memoirs of the Geological Survey of Great Britain, England and Wales, HMSO.

Proakis, J.G. & Manolakis, D.K. (2006). Digital Signal Processing (4th Edition), Prentice Hall. ISBN 978-01-318-7374-2

Riley, K.F. (1974 [and subsequent editions]). *Mathematical Methods for the Physical Sciences: An Informal Treatment for Students of Physics and Engineering,* Cambridge University Press, ISBN 978-05-210-9839-7.

Rilling, G., Flandrin, P. & Goncalves, P. (2003). On Empirical Mode Decomposition and its Algorithms, *IEEE-EURASIP Workshop on Nonlinear Signal and Image Processing NSIP-03,* Grado (I)

Smith, K.A., Gillmore, G.K. & Sinclair, J.M. (2000). Sediments and Ostracoda from Courteenhall, Northamptonshire, U.K. and their implications for the depositional environment of the Pleistocene Milton Formation, *Proc. Geologists' Association,* Vol.111, No.3, pp.253-263, doi:10.1016/S0016-7878(00)80018-8.

Toghill, P. (2003). *The Geology of Britain: an Introduction,* Airlife, Wiltshire, UK. ISBN 1-84037-404-7.

Venables, W.N. & Ripley, B.D. (2002). *Modern Applied Statistics with S.* (4th Edition), Springer. ISBN 0-387-95457-0 and online complement http://www.stats.ox.ac.uk/pub/MASS4/

Wakita, H. (1996). Geochemical Challenge to Earthquake Prediction, *Proc. USA National Academy Science,* Vol.93, No.9, pp.3781-3786.

Walia, V., Virk, H.S., Yang, T.F., Mahajan, S., Walia, M. & Bajwa, B.S. (2005). Earthquake prediction studies using radon as a precursor in N-W Himalayas, India: A case study. *Terrestrial, Atmospheric and Oceanic Sciences,* Vol.16, No.4, pp.775-804. ISSN 1017-0839.

Walia, V., Virk, H.S. & Bajwa, B.S. (2006). Radon precursory signals for some earthquakes of magnitude > 5 occurred in N-W Himalaya. *Pure and Applied Geophysics*, Vol.163, No.4, pp.711-721. ISSN 0033-4553.

Zmazek, B., Vaupotic, J., Zivcic, M., Premru, U. & Kobal, I. (2000). Radon Measurements for Earthquake Prediction in Slovenia, *Fizika B*, Vol.9, No.3, pp.111-118. ISSN 1330-0016 / 1333-9133.

6

Earthquake Prediction: Analogy with Forecasting Models for Cyber Attacks in Internet and Computer Systems

Elvis Pontes, Anderson A. A. Silva, Adilson E. Guelfi and Sérgio T. Kofuji
Laboratory of Integrated Systems, Polytechnic School of the University of São Paulo
Brazil

1. Introduction

Currently, security of the cyber space (computer networks and the Internet) is mostly based on detection and/or blocking of attacks by the use of Intrusion Detection and Prevention System (IDPS), according to (National Institute of Standards and Technology [NIST SP800-94], 2010). However IDPS lacks in security as it is based on *postmortem* approaches - threats and attacks are identified and/or blocked only after they can inflict serious damage to the computer systems either while attacks are happening, or when attacks have already imposed losses to the systems (Haslum et al, 2008).

On the subject of earthquakes, one can notice the same kind of limitation: once an earthquake has already begun, devices can provide warnings with just few seconds before major shaking arrives at a given location (Bleier & Freund, 2005), (Su & Zhu, 2009). In the cyber space context, intending to cover the deficiency of late warnings, predicting techniques have already been approached in a small number of studies for cyber attacks in the last few years (Pontes & Zucchi, 2010), (Haslum et al, 2008), (Lai-Chenq, 2007), (Yin et al 2004).

1.1 Motivation

Although studies based on 1) historical earthquake records and 2) monitoring the earth's surface had contributed to map affected regions, short-term earthquake predictions are not efficient yet (Bleier & Freund, 2005).

Some researchers are studying and correlating signals gathered in the ionosphere that can precede earthquakes, like odd radio noise and lights in the sky.

According to (Bleier & Freund, 2005) "both the lights and the radio waves appear to be electromagnetic disturbances that happen when crystalline rocks are deformed--or even broken--by the slow grinding of the earth that occurs just before the dramatic slip that is an earthquake".

Some occurrences of earthquakes show signals and disturbances like following reported ones:

- Loma Prieta, San Francisco,1989: two weeks before a 7.1-magnitude earthquake, strong signals (20 times that of normal background noise at the 0.01 Hz frequency) of magnetic disturbance were detected. Three times before the quake the signals jumped to 60 times normal size at the 0.01 Hz frequency;

- Spitak, Armenia, 1988: signals occurred shortly before a 6.9-magnitude quake;
- Guam, Pacific Ocean, 1993: signals were observed before a 8.0-magnitude quake;
- Parkfield, California, 2003: nine hours before a 6.0-magnitude quake, spikes of activity, four to five times normal size (0.2 to 0.9 Hz frequency) were detected;
- Taiwan, 1999: sensors registered unusually large disturbance in a normally quiet signal before the 7.7-magnitude earthquake. Researchers calculated the current required to generate those magnetic-field disturbance: between 1 million and 100 million amperes.

Those examples show that the occurrence of electromagnetic signals does not justify a public warning, but it is an important source of data for forecasters and are also useful for directing the course of research on earthquake prediction such as changes in the conductivity of the air over the quake zone caused by current welling up from the ground, that contribute to the formation of the so-called earthquake lights in the Mojave Desert (Fig. 1).

Fig. 1. Earthquake lights (Bleier & Freund, 2005)

There are some theories about these signals generation, but details are not conclusive yet. Notwithstanding, electromagnetic effects of the signals can be detected in a number of ways (see Fig. 2 next page).

Ground-based sensors, monitor changes in the low-frequency magnetic field and measure changes in the conductivity level of the air. Satellites monitor noise level at extremely low frequency and monitor the infrared light which is probably emitted when rocks are deformed or even broken. Some example:

- after the 1989 Armenia earthquake, electromagnetic Extremely Lower Frequency (ELF) disturbances were observed by a Soviet Cosmos satellite by a month;
- an U.S. satellite detected ELF bursts before and after a 6.5-magnitude earthquake in 2003 at California;
- In 2004 France has launched a satellite for Detection of Electro-Magnetic Emissions Transmitted from Earthquake Regions (DEMETER) that unfortunately presented malfunctioning.

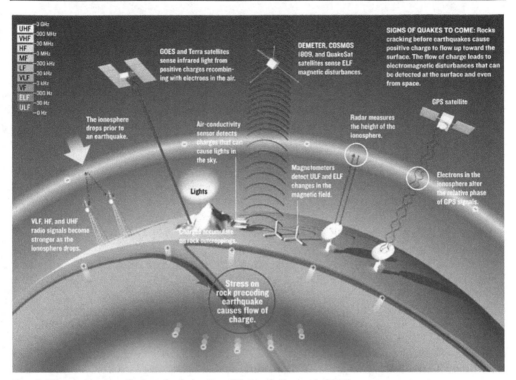

Fig. 2. Electromagnetic signals detection (Bleier & Freund, 2005)

According to (Bleier & Freund, 2005), "infrared radiation detected by satellites may also prove to be a warning sign of earthquakes to come". In China satellite-based instruments had registered the occurrence of several infrared signature instances with a jump of 4 to 5 oC before some earthquakes during the past two decades Sensors in NASA's Terra Earth Observing System satellite registered what NASA called a thermal anomaly on 21 January 2001 in Gujarat, India, just five days before a 7.7-magnitude quake there; the anomaly was gone a few days after the quake (Fig. 3). Accordingly with (Bleier & Freund, 2005), in both cases researches believe these sensosrs have detected an infrared luminescence generated by the recombination of electrons and holes, not a real temperature increase.

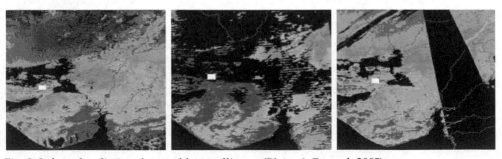

Fig. 3. Infrared radiation detected by satellites n (Bleier & Freund, 2005)

The connection between large earthquakes and electromagnetic phenomena in the ground and in the ionosphere is becoming increasingly solid. Researchers in many countries, including China, France, Greece, Italy, Japan, Taiwan, and the United States, are now contributing to the data by monitoring known earthquake zones.

Some correlations between historical data can be traced as well: monitoring 144 earthquakes (1997-1999), Taiwanese researches noticed significant changes in the electron content of the ionosphere some days before the quakes higher than 6-magnitude.

Therefore, the integration of: (1) several types of sensors (ground and space-based), (2) a network to bring together those signals, (3) a good distribution of the sensors (several sensors in a large area), (4) several types of detection (Ultral Low Frequency (ULF), ELF and magnetic-field changes, ionospheric changes, infrared luminescence, and air-conductivity changes--along with traditional mechanical and GPS monitoring of movements of the earth's crust and (5) the correlation of all data gathered, could make forecast more reliable.

1.2 Analogy with forecasting in cyber security

Cyber attacks can be classified as a set of actions with the purpose of compromising the integrity, confidentiality or availability of computer systems. Cyber attacks can be caused by users or malicious software, which try either to obtain access, to use systems in an unauthorized way, or to enumerate privileges (NIST SP800-94, 2010).

(Internet Crime Complaint Center [IC3], 2010) published a study in the United States about losses in 2009 concerning cyber-attacks: frauds in cyber space caused about $559.7 million of losses in 336,655 organizations. This was a 111,5% increase for the losses and a 22.3% increase for the complaints, as compared to 2008 when 275,284 complaints were received, reporting $264.6 million in total losses. According to (McPherson & Labovitz, 2010), in 2009 the largest reported volumetric Distributed Denial of Service (DDoS) attack exceeded 49 Gbps sustained towards a single target in Europe.

Beyond sheer attack size, (McPherson & Labovitz, 2010) indicated that cyber-attacks become more sophisticated, with attackers expressly aiming to exhaust resources other than bandwidth, such as firewalls, load-balancers, back-end database infrastructure and associated transaction capacity, cached data serving algorithms, etc. This increasing sophistication is a trend that has been captured in previous editions of the survey of (McPherson & Labovitz, 2010) as well. Regarding DDoS attacks, it is expected these attacks to become more common against independent media and human rights sites in 2011, as the recent highly publicized DDoS attacks on Wikileaks, and "Operation Payback" attacks by "Anonymous" on sites perceived to oppose Wikileaks (Zuckerman et al, 2010).

According to (Pontes et al, 2008), (Pontes & Guelfi, 2009a), (Pontes & Guelfi, 2009b), (Pontes & Zucchi, 2010), an early warning system showing a future trend outlook with an increasing number of cyber-attacks, exposed by forecasting analysis, may influence decisions on the security devices adoption (e.g. rules in IDPS combined with rules in firewalls) before incidents happen, according to the needs. Although, three major gaps lie in the studies about forecasting of cyber attacks: a) the use of few sensors and/or sensors employed locally; b) the use of just one forecasting technique; and c) lack of information sharing among sensors to be used for correlation (Pontes & Guelfi, 2009a). Correlation of information between IDPS and forecasters means looking for similar characteristics that may be related (Pontes & Guelfi, 2009a) (Abad et al, 2003). Throughout correlation it is possible to eliminate redundant and false data, to discover attack patterns and understand attack strategies (Zhay et al, 2006).

Nevertheless, forecasts and alert correlation may be challenging as they depend on the reliability of the source of the security alerts (Silva & Guelfi, 2010). Therefore, the precision level of the detection tools is an important issue for validating correlations. Multi-correlation or integration of alerts with information from different sources, e.g. tools for monitoring or operating system logs, can allow a new classification for alerts, improving accuracy of the results (Abad et al, 2003), (Zhay et al, 2006). References (Abad et al, 2003), (Zhay et al, 2006), (Zhay et al, 2004) employed multi-correlation; however neither a detailed analysis concerning influence of isolated alerts in the FP rates, nor forecasting techniques were not applied for predicting future attacks (forecasting).

Forecasting analysis in the information security area can be similar to forecasting methodologies used in any other fields: meteorology, for instance, use sensors to capture data about temperature, humidity, etc (Lajara et al, 2007), (Lorenz, 2005); seismology employs sensors to capture electromagnetic emissions from the rocks (Bleier & Freund, 2005); for economics, specifically stock market, data is collected from diverse companies (annual profit, potential customers, assets, etc) to draw trends about shares of companies (Prechter & Frost, 2002), (Mandelbrot & Hudson, 2006). For any field formal models can be applied to predict events over the collected data. But, before applying formal models, data regarding different kind of variables should be correlated (Armstrong, 2002). According to (Armstrong, 2002), to obtain a more accurate and realistic result about predictions it is suggested: (1) to use diverse forecasting techniques; (2) to analyze information regarding diverse variables and acquired data, from sensors for instance; (3) to employ diverse kind of employed forecasting models.

Concerning cyber attacks, (Lai-Chenq, 2007), (Yin et al 2004) employed forecasting models, however they used just one formal method for predicting events and they did not make use of any kind of correlation process. In this chapter, security events for cyber security are actions, processes that have an effect on the system, disregarding the kind of the effect – in other words, actions that could result in positive or negative effects on the system. In other hand, security alerts are types of security events, indicating anomalous activities or cyber attacks (Silva & Guelfi, 2010). In our earlier works we proposed the Distributed Intrusion Forecasting System (DIFS) (Pontes & Guelfi, 2009), (Pontes & Zucchi, 2010), which covered the following gaps of today's forecasting techniques in IDPS: a) the use of few sensors and/or sensors employed locally for capturing data; b) the use of just one forecasting technique; and c) lack of information sharing among sensors to be used for correlation. Notwithstanding, we faced huge amount of alerts which could have negative influence over forecasting results.

1.3 Proposal
The goal of this chapter is to propose a Distributed Intrusion Forecasting System (DIFS) with a two stage system which allows: (1) in the first stage it is possible to make a correlation of security alerts using an Event Analysis System (EAS); and (2) to apply forecasting techniques on the data (historical series) generated by the previous stage (EAS). The DIFS works with prediction models and sensors acting in different network levels (host, border and backbone), which enables the use of different forecasting techniques (e.g. Fibonacci sequence and moving averages), the cooperation among points of analysis and the correlation of predictions. Additionally to the main goal, the aim of this chapter is proposing an analogous approach for earthquake prediction. As results it is intended to increase reliability of incidents predictions (e.g. earthquake incidents, cyber attacks), to prevent

incidents in a proactive manner and to improve risk management employed for security of the homeland cyber space. A proof of concept of such architecture (DIFS) is presented, which allows concluding about the improvement of forecasts in the cyber space; furthermore, tests applied over two datasets - (Defense Advanced Research Projects Agency [DARPA], 1998) and (Knowledge Discovery and Data Mining Tools Competition [KDD], 1999) - with an IDPS have shown that the employed techniques define incidents trends.

This chapter is organized as follows: state of art concerning forecasting and event correlation in IDPS are in section 2. Section 3 introduces the proposal of this chapter: the DIFS and the two stage system for correlation regarding cyber attacks. Section 4 presents details about the tests and environment to validate the proposal. Results are analyzed in section 5 and section 6 summarizes conclusions and suggestions for new studies.

2. State of art – Cyber attacks, event correlation and forecasting

In this section we approach event correlation for detecting cyber-attacks, the forecasting methods used to predict cyber-attacks and Distributed Architecture for Intrusion Forecasting System (DIFS (Pontes & Guelfi, 2009), (Pontes & Zucchi, 2010).

2.1 Unwanted internet traffic and cyber attacks

The expression "unwanted traffic" was first introduced in the eighties and it has always been related to malicious activity as worms, virus, intrusions etc (Feitosa et al, 2008). Reference (Feitosa et al, 2008) defines unwanted Internet traffic (UIT) as unproductive and useless traffic, with malicious (worms, scans, spam) and benign (wrong setting in the routers) events. Reference (Soto, 2005) completes this definition: UIT may result from the noise in the telecommunication network. (Andersson et al, 2007) classified UIT as the malicious or useless one, with the objective to compromise vulnerable hosts, to spread malicious code, spam, DoS and DDoS. UIT may also be junk traffic, background traffic and anomalous traffic.

Symposiums and workshops have been done about the issue of UIT, like the one promoted by Internet Architecture Board (IAB), on March 2006 (Andersson et al, 2007) and April 2008: the intention was to share information among people from different fields and organizations, fostering an interchange of experiences, views, and ideas between the various research communities. As a result, the Request for Comments (RFC) 4948 details the UIT types, the main causes, existent solutions and the actions to be taken in short and long term. It was decided, in this workshop, that some other research topics about UIT would be managed by the IAB, Internet Engineering Task Force (IETF) and Internet Research Task Force (IRTF).

According to (Feitosa et al, 2008), several of the losses caused by UIT are due to the inefficiency of today's techniques and security devices (anti-spam, antivirus, Intrusion Detection and Prevention Systems (IDPS) (NIST, 2010), firewalls), whether for detecting and preventing the intrusion, or to treat the UIT. Furthermore, the high rates of false positives, false negatives and the lack of a forecasting approach for the Internet traffic are some of the reasons of the UIT increasing. Internet attacks continue apace, with UIT, such as phishing, spam, and distributed denial of service attacks increasing steadily. However, it is important to classify whether it is unwanted or not: Voip (Skype), peer-to-peer (P2P), instant messengers (MSN, Google talk, ICQ), online social networks. Different classification may be employed from one company to another, from user to user, from country to country. China, for instance, does not allow calls from Skype to telephones. Another example: routers for backbone providers and for small companies - the UIT is differently classified in both cases (Feitosa et al, 2008).

2.2 Approaches for correlation of security events

Correlation techniques for security events can be classified into three categories: (1) rule-based, (2) based on anomaly and (3) based on causes and consequences (Prerequisites and Consequences (PC)) (Abad et al, 2003). The rule-based method requires some prior knowledge about the attack, so the target machine has to pass through a preparation phase called training. The goal of this phase is to make the target machine able to precisely detect the vulnerabilities in which the target machine was trained for (Abad et al, 2003), (Mizoguchi, 2000). Gaps of rule-based method are: (1) it is computer intensive; (2) it results in lots of data; (3) the method works only for known vulnerabilities.

The method based on anomaly analyzes network data flow, using correlation with statistical methods, using accumulation of gathered information and using observations of the occurred deviations throughout processes of network data flow; in a manner to allow detecting new attacks. For instance, (Manikopoulos & Papavassiliou, 2002) demonstrates a system for detecting anomalies which is characterized by monitoring several parameters simultaneously. Reference (Valdes & Skinner, 2001) presents a probabilistic correlation proposed for IDPS, based on data fusion and multi-sensors. However, the method which uses anomaly cannot detect anomalous activity hidden in a normal process, if it is performed at very low levels. Besides, as this method analyzes normal processes reporting only wrong deviations, hence the method is not suitable for finding causes of attacks (Ning et al, 2001).

The PC method lies on connections between causes (conditions for an attack to be true) and consequences (results of the exploitation of a cause), in order to correlate alerts based on the gathered information. This method is suitable for discovering strategies of attacks. Both causes and consequences are composed of information concerning attributes of alerts (specific features belonging to each alert) and are correlated. Arrangement of attributes is called tuple. According to Fig. 4, for the connections to be valid, a preparatory alert must have in its consequences at least one tuple, which repeats in the causes of the resulting alert. In other words, the preparatory alert contributes to the construction of the resulting alert, and therefore it can be correlated. For this connection, illustrated by Fig. 4, the timestamp of the preparatory alert has to come before the resulting alert (Silva & Guelfi, 2010), (Pontes & Guelfi, 2010), (Ning et al, 2001).

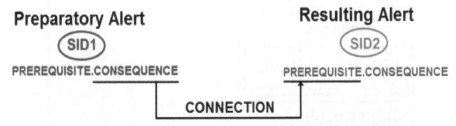

Fig. 4. Connections Between Alerts - Consequence of Preparatory Alert (SID1) is Connected to Prerequisites of Resulting Alert (SID2).

In order to reduce complexity, correlation can be shown in graphs where alerts are represented by nodes and connections are depicted by arrows (representing correlations between alerts).

Yet, some gaps in the PC method may be mentioned, such as the difficulty in obtaining causes and consequences of alerts (Pietraszek & Tanner, 2005), the impossibility to analyze isolated alerts (alerts that are not correlated) and the fact that missed attacks are hard to correlate. An alternative to minimize the problem is to apply complementary correlation techniques (Morin & Debar, 2003), using sensors to work in cooperation, in order to supervise the environment for minimizing missed detections. There are two techniques to map IDPS' alerts and logs obtained from other sources: descending analysis and ascending analysis (Abad et al, 2003), (Silva, 2010).

Descending analysis is based on the investigation of occurred attacks, verifying (correlating) whether other logs (e.g. logs from O.S.) have or do not have vestiges of the attacks' incident. For occurred attack, other traced logs (e.g. Operational System's logs) can be analyzed based on timestamp. This type of analysis is useful to trace evidences about strategies of events, in order to map attacks to its source.

The ascending technique is used to discover attacks by the analysis of several logs. Once an anomaly is detected in one of these logs, other logs are checked based on timestamp. Although ascending technique is computer intensive, this technique allows detecting new attacks.

In an earlier work we proposed the EAS (Silva & Guelfi, 2010), (Silva, 2010), intending to improve results of security events correlation and intrusion detection. EAS is able to make multi-correlation for events from Operational Systems (OSs) and from IDPS (log analysis), consequently, EAS is also capable for verifying the influence of isolated alerts in the cyber-security context.

The EAS architecture has 4 modules, as shown by Fig.5: (a) converter: the aim of this module is to handle the input data into the system (IDPS signatures, alerts and logs from the OS); (b) updating: it controls data which is going to be used by the system; (c) correlating: it does mappings for the correlation processes, FP identification, and the identification of isolated alerts; (d) calculator: it analyzes and compares FP, based on the results from the correlating module.

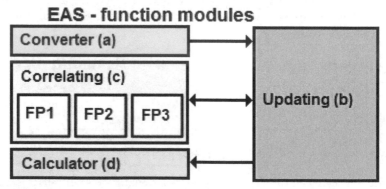

Fig. 5. EAS's Architecture (Silva, 2010), (Silva & Guelfi, 2010)

According to (Silva, 2010), (Silva & Guelfi, 2010), with the employment of the EAS it was possible to improve the today's results of correlation regarding security events, considering the following issues: (1) traceability for causes and consequences within the PC-correlation method (with multi-correlation criteria, correlation analysis (ascending/descending) and identification of FP alerts through tables and graphs); and (2) the process of results validation regarding the correlation. In (Silva, 2010), (Silva & Guelfi, 2010), results of correlating phase were evaluated in three steps (FP1, FP2 and FP3) using tables and graphs. The stepwise analysis allowed comparison of the results. EAS achieved an increase of 112.09% in the identification of FP alerts after the multi-correlation. Another important result of EAS was the evidence of preparatory connections between individual alerts that are in fact part of larger and more elaborated attacks. In other words, EAS can show that individual alerts can be grouped in a single attack, since they are part of the same attack strategy (Silva, 2010), (Silva & Guelfi, 2010).

2.3 Related forecasting methodologies for earthquakes

Statistical based forecast methodologies are used to understand and predict earthquake signals (Kagan, & Jackson, 2000). It is important to discuss these other researches to notice the variety of forecasting applications. Two forecast researches are summarized below.

2.3.1 Earthquake forecasting and its verification

Holliday et. al (2005) has based their forecast research on the association of occurrence of small earthquakes with probably future large ones. In fact, the method does not predict earthquakes, but spots regions (Hotspots regions) where they are most likely to occur in the future (about ten years).

Basically the research objective is to reduce risk areas analyzing the historical seismicity for anomalous behaviour.

The approach is based on a pattern informatics (PI) method which quantifies temporal variations in seismicity and is as follows Holliday et al, (2005):

1. The region of interest is divided into N_B square boxes with linear dimension Δx. Boxes are identified by a subscript i and are centered at x_i. For each box, there is a time series $N_i(t)$, which is the number of earthquakes per unit time at time t larger than the lower cut-off magnitude Mc. The time series in box i is defined between a base time t_b and the present time t.

2. All earthquakes in the region of interest with magnitudes greater than a lower cutoff magnitude M_c are included. The lower cutoff magnitude M_c is specified in order to ensure completeness of the data through time, from an initial time t_0 to a final time t_2.

3. Three time intervals are considered:

a. A reference time interval from t_b to t_1.

b. A second time interval from t_b to t_2, $t_2 > t_1$. The change interval over which seismic activity changes are determined is then $t_2 - t_1$. The time t_b is chosen to lie between t0 and t1. Typically we take $t_0 = 1932$, $t_1 = 1990$, and $t_2 = 2000$. The objective is to quantify anomalous seismic activity in the change interval t_2 to t_1 relative to the reference interval t_b to t_1.

c. The forecast time interval t_2 to t_3, for which the forecast is valid. The change and forecast intervals are taken and forecast intervals to have the same length. For the above example, $t_3 = 2010$.

4. The seismic intensity in box i, $I_i(t_b, t)$, between two times $t_b < t$, can then be defined as the average number of earthquakes with magnitudes greater than M_c that occur in the box per unit time during the specified time interval t_b to t. Therefore, using discrete notation, we can write:

$$I_i(t_b, t) = \frac{1}{t - t_b} \sum_{t'=t_b}^{t} N_i(t').$$ (1)

Where the sum is performed over increments of the time series, say days.

5. In order to compare the intensities from different time intervals, it is required that they have the same statistical properties. Therfore, the seismic intensities are normalized by subtracting the mean seismic activity of all boxes and dividing by the standard deviation of the seismic activity in all boxes. The statistically normalized seismic intensity of box i during the time interval t_b to t is then defined by

$$\hat{I}_i(t_b, t) = \frac{I_i(t_b, t) - < I_i(t_b, t) >}{\sigma(t_b, t)},$$ (2)

Where $< I_i(t_b, t) >$ is the mean intensity averaged over all the boxes and $\sigma(t_b, t)$ is the standard deviation of intensity over all the boxes.

6. The measure of anomalous seismicity in box i is the difference between the two normalized seismic intensities:

$$\Delta I_i(t_b, t_1, t_2) = \hat{I}_i(t_b, t_2) - \hat{I}_i(t_b, t_1).$$ (3)

7. To reduce the relative importance of random fluctuations (noise) in seismic activity, the average change in intensity is computed, $\underline{\Delta I_i(t_0, t_1, t_2)}$ over all possible pairs of normalized intensity maps having the same change interval:

$$\underline{\Delta I_i(t_0, t_1, t_2)} = \frac{1}{t_1 - t_0} \sum_{t_b=t_0}^{t_1} \Delta I_i(t_b, t_1, t_2).$$ (4)

Where the sum is performed over increments of the time series, which here are days.

8. The probability is defined as a future earthquake in box i, $P_i(t_0, t_1, t_2,)$, as the square of the average intensity change:

$$P_i(t_0, t_1, t_2,) = \underline{\Delta I_i(t_b, t_1, t_2)}^2.$$ (5)

9. To identify anomalous regions, it is desirable to compute the change in the probability $P_i(t_0, t_1, t_2,)$ relative to the background so that we subtract the mean probability over all boxes. This change in the probability is denoted by

$$\Delta P_i(t_0, t_1, t_2) = P_i(t_0, t_1, t_2) - < P_i(t_0, t_1, t_2) >,$$ (6)

Where $< P (t, t, t) >$ is the background probability hotspots are defined to be the regions where $\Delta P_i(t_0, t_1, t_2)$ is positive. In these regions, $P_i(t_0, t_1, t_2)$ is larger than the average value for all boxes (the background level). Note that since the intensities are squared in defining probabilities the hotspots may be due to either increases of seismic activity during the change time interval (activation) or due to decreases (quiescence). The following hypothesis is taken into account: earthquakes with magnitudes larger than $M_c + 2$ will occur preferentially in hotspots during the forecast time interval t_2 to t_3.

To evaluate the model a Relative Operating Characteristic (ROC) diagram, which can be viewed as binary forecast either to occur or not to occur, was used and presented significant results with a relative high proportion of hotspots representing locations of probably future large earthquakes.

Although good results, the model Holliday, J.R., et. al, (2005) could be used as an input in a larger forecast system like DIFSA which would provide the communication and correlation of data with others different models.

2.3.2 Probabilistic forecasting of earthquakes

(Kagan, & Jackson, 2000) has developed a research with both short and long-term forecast approach and testing both with a likelihood function to 5.8-magnitude (or larger) quakes. Although the long-term approach (see Table 1), is not completely developed and is suitable to estimation of occurrence of earthquakes, it is derived from statistical, physical and intuitive arguments while the short-term forecast seismicity model is based on a specific stochastic model and updated daily (see Table 1).

The research assumes that the rate density (probability per unit area and time) is proportional to a smoothed version of past seismicity and depends approximately on a negative power of the epicentral distance and linearly on magnitude of the past earthquakes.

The model (Kagan, & Jackson, 2000) does not use retrospective evaluation of seismic data. The parameters of long-term are evaluated on the basis of success in the forecasting of seismic activity also indicating possible earthquakes perturbations. A maximum likelihood procedure to infer optimal values are applied on short-term approach which can be incorporated into real-time seismic networks to provide seismic hazard estimate.

About the scientific results (Kagan, & Jackson, 2000) concluded that the research depicted a statistical relationship between successive earthquakes in a quantitative way that facilitate hypothesis testing. About the practical results the quantitative predictive assessment can be adopted into mitigation strategies.

Latitude	Longitude	Long-term forecast						Short-term forecast	
		Probability $m \geq 5.8$ eq/day*km^2	\multicolumn{5}{c}{Focal mechanism}		Probability $m \geq 5.8$ eq/day*km^2 time-dependent	Probability ratio time-dependent/ independent
			T-axis		P-axis		Rotation angle degree		
			Pl	Az	Pl	Az			
119.5	19.5	3.18E-09	31	208	10	304	64.8	1.79E-14	5.62E-06
120.0	19.5	5.23E-09	17	213	32	314	68.8	1.41E-10	2.71E-02
120.5	19.5	4.28E-08	7	93	75	335	21.4	2.12E-07	5.0
121.0	19.5	3.02E-08	69	135	21	302	28.2	2.84E-07	9.4
121.5	19.5	1.82E-08	77	106	13	296	40.9	6.14E-08	3.4
122.0	19.5	7.81E-09	60	32	3	297	48.4	1.13E-10	1.45E-02
122.5	19.5	4.15E-09	81	228	4	113	51.8	1.00E-12	2.41E-04
123.0	19.5	3.01E-09	78	251	9	110	50.3	7.70E-16	2.56E-07
123.5	19.5	2.43E-09	76	273	13	107	49.5	1.08E-20	4.43E-12

Table 1. Example of long- and short-term forecast, 1999 February 11, north of Philippines.(Kagan, & Jackson, 2000)

The versatility of the methodology based on forecasts is evident in this work, presenting significant results. This scenario shows that quite different methods (e.g, that use and do not use historical data) can be used in conjunction with an approach that uses DIFSA.

2.4 Forecasting for cyber attacks

The forecasting approaches in IDPS lie mainly on stochastic methods (Ramasubramanian & Kannan, 2004), (Alampalayam & Kumar, 2004), (Chung et al, 2006). With no attention about predictions, references (Ye et al, 2001), (Ye et al, 2003), (Wong et al, 2006) applied diverse probabilistic techniques (decision tree, Hotelling's T^2 test, chi-square multivariate, Markov chain and Exponential Weighted Moving Average (EWMA)) on audit data as a way to analyze three properties of the UIT: frequency, duration, and ordering. Reference (Ye et al, 2001), (Ye et al, 2003) has come to the following findings: 1) The sequence of events is necessary for IDPS, as a single audit event at a given time is not sufficient; 2) Ordering (transaction (Wong et al, 2006)) provides additional advantage to the frequency property, but it is computationally intensive. According to (Ye et al, 2001), (Ye et al, 2003), (Wong et al, 2006), the frequency property by itself provides good intrusion detection. References (Ye et al, 2001), (Ye et al, 2003), (Wong et al, 2006) did not approach correlation for IDPS.

Moving averages (simple, weighted, EWMA, or central) with time series data are regularly used to smooth out fluctuations and highlight trends (NIST, 2009). EWMA may be applied for auto correlated and uncorrelated data for detecting cyber-attacks which manifest themselves through significant changes in the intensity of events occurring (Ye et al, 2001). Both (EWMA for auto correlated and uncorrelated) has presented good efficiency for detecting attacks. EWMA applies weighting factors which decrease, giving much more importance to recent observations while still not discarding older observations entirely. The statistic that is calculated is (NIST, 2009):

$$EWMA_t = \alpha Y_t + (1 - \alpha)EWMA_{t-1} \quad \text{for t=1, 2, ..., n.} \tag{7}$$

Where: EWMA is the mean of historical data; Yt is the observation at time t; n is the number of observations to be monitored including EWMA; $0 < \alpha < 1$ is a constant that determines the depth of memory of the EWMA.

The parameter α determines the rate of weight of older data into the calculation of the EWMA statistic. So, a large value of α gives more weight to recent data and less weight to older data; a small value of α gives more weight to older data.

Reference (Cisar and Cisar, 2007) gives an overview of adopting EWMA with adaptive thresholds, based on normal profile of network traffic. The analysis of thresholds with EWMA may summarize huge amount of data in network traffic (Zhay et al, 2006), (Pontes & Zucchi, 2010). Diverse moving averages, combined with Fibonacci sequence forecasting approach, were also used by (Zuckerman et al, 2010) to spot trends of cyber attacks in the (DARPA, 1998) datasets.

A simple moving average (SMA) is the non weighted mean of the previous n data. For example, a 10-hours SMA of intrusive event X (DoS, e.g.) is the mean of the previous 10 hours' event X. If those events are: $e_M, e_{M-1}, ..., e_{M-9}$. Then the formula is (NIST, 2009), (Roberts, 1959):

$$SMA = \frac{e_M + e_{M-1} + \cdots + e_{M-9}}{10} \tag{8}$$

When calculating successive values, a new value comes into the sum and an old value drops out, meaning a full summation each time is unnecessary,

$$SMA_{current\ hour} = SMA_{last\ hour} - \frac{e_M - n}{n} + \frac{e_M}{n} \tag{9}$$

Nevertheless, the forecasting approaches which use moving averages to cope with cyber attacks in IDPS are limited to analyze cyber attacks individually, e.g. in just one IDPS. Therefore, there is no collaboration among the forecasters. Besides: the concept of sensors is not adopted in (Pontes et al, 2008), (Pontes & Guelfi, 2009a), (Pontes & Guelfi, 2009b), (Pontes & Zucchi, 2010), (Ishida et al, 2005), (Viinikka et al, 2006), (Ye et al, 2003).

3. The distributed intrusion forecasting system with the two stage system (Pontes et al, 2011)

Intrusion Forecasting Systems (IFS) can work proactively in cyber security contexts, as early warning systems, in order to indicate or identify UIT (incidents, threats, attacks) in advance. IFS can also represent an improvement of IDPS, which is based on postmortem approaches (UIT is identified and/or blocked only after they can inflict serious damage to the computer systems). IFS predicts UIT by the use of different forecasting techniques (for instance, moving average, Fibonacci sequence etc) applied either for local or distributed environment. Additionally, for distributed environments, e.g. DIFS, the use of cooperative sensors can improve accuracy about predictions of incidents.

Fig. 6 depicts the proposal of this chapter, i.e. the DIFS and the forecasting levels. Similarly to forecasting methodologies used in other fields (e.g. Meteorology), DIFS also spreads agents and/or sensors widely to make predictions about the different kinds of UIT (spam, virus, intrusion, abnormal network traffic). There are four levels of the IFS: level 1 - independent security devices of hosts; level 2 - integrated security devices of hosts; level 3 - the network level; and level 4 - the backbone level. All levels have some communication degree among each other. In other words, the forecasts obtained from level 1 are shared and correlated to the forecasts of the other levels. Lower levels work as sensors to higher levels; consequently feedback about the UIT trends may be exchanged from one level to another.

Level 1 concerns the trend analysis about incidents, alerts and diagnosis reported independently by the hosts' security devices (antivirus, antispyware, host-based IDPS and other anomaly detector systems). For each security device, individual forecasts may be provided, e.g. the trend about spam for next hour or the day of tomorrow, or the trend about virus infection etc. The next step of the IFS level 1 is to help the hosts' security devices to determine whether or not they should adopt countermeasures to stop UIT

Level 2 involves correlation of forecasts about the hosts' security devices. At this level, the analysis lays on two databases: a) All the historical data generated from each one of the hosts' security devices are processed individually by the IFS first level, then stored in a database; b) The network flow may also be recorded for further forecasting analysis. The next step for the IFS level 2 is to query and to analyze the trends (forecasts) of such databases. After analyzing it, IFS level 2 returns a feedback to IFS level 1. It is important to notice that the databases of IFS level 1 work as sensors for IFS level 2.

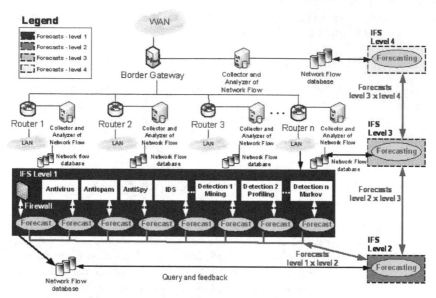

Fig. 6. DIFS Architecture - adapted from (Pontes & Guelfi, 2009)

The implementation of IFS level 3 happens at the gateway of the LAN. IFS level 3 is analogous to IFS level 2, as it queries databases generated by IFS levels 1 and 2. Likewise IFS level 1, some security devices may be installed at the gateway (as firewall, regular IDPS, etc) and they may also be analyzed. The steps for analysis at this level are: a) Network security devices record UIT in databases; IFS level 3 queries the databases provided by the lower levels and current level; b) IFS level 3 analyzes the provided databases to define trends; c) IFS level 3 provides feedback of the trend analysis to the security devices; d) IFS level 3 may also give feedback for the lower levels. It is important to notice that IFS level 1 and level 2 databases work as sensors for IFS level 3. The sensor elements may be more numerous at IFS level 3.

IFS level 4 is the major level. It considers the structure of the backbone providers (an ISP, for instance). In the same way IFS level 3 and level 2, different security devices are linked to the backbone level. The steps for IFS level 4 to work are: a) Backbone security devices record UIT in database; b) IFS level 4 queries the databases provided by the lower and current level; c) IFS level 4 analyzes the provided databases to define the trends; d) IFS level 4 provides feedback of the trend analysis to the current level; e) IFS level 4 may also give feedback for the lower levels. Similarly to lower levels, IFS level 4 uses the same concept of sensors: lower databases and the entire lower IFS levels are sensors for IFS level 4. An important note is: the IFS level 4 may be shared and correlated among various backbone providers. To correlate forecasts of IFS level 4 means to provide the most realistic and integrated trend about UIT, as it may spread sensors along the network (Lajara et al, 2007).

It is important to notice that for the IFS we implemented a two stage system (Pontes et al, 2011), intending to improve the forecasting results by the use of correlation. Fig. 7 presents the sequence of activities done by the system:

1. The first task is the multi-correlation, running the EAS, to filter FP and tracing sophisticated. During this step, OS's logs, IDPS's logs, network traffic and other logs are analyzed by the EAS. According to Fig. 4, diverse logs and network traffic represent the Entry 1 for the two stage system.

2. The second task is done by the IFS, applying forecasting techniques over the EAS' generated data (historical series, without a considerable amount of FP). Several forecasting techniques may be adopted in this stage (e.g. EWMA, Fibonacci sequence, Markov chains). As illustrated by Fig. 7, EAS' generated data is the Entry 2 for the two stage system. Sep 2 of the two stage system considers just data from Entry 2.

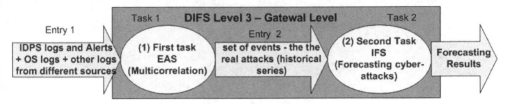

Fig. 7. Sequence of Steps: (1) EAS Filtering – (2) IFS (Pontes et al, 2011)

4. Proof of concept

In this section we are going to describe two of the prototypes we have prepared and analysed. In the first one (Pontes et al, 2009), for the proof of concept, levels 1, 2 and 3 of the DIFS were implemented in three sites geographically divided (A, A' and A''). The following hardware and services were used: a) 1 Pentium core 2 quad 2.0 GHz, 8GB RAM; b) 2 Pentium core 2 duo 1.8 GHz, 4GB RAM; c) 10 virtual machines (Ubuntu 8.04) 512MB RAM; d) 4 virtual machines (WindowsXP) 512MB RAM; e) Windows Vista (host for the virtual machines); VMware Player 2.51; Snort; Netfilter/Iptables; MySQL; OpenVPN.

Likewise (Haslum et al, 2008), in this prototype the simulation of UIT was divided in just in four types: 1) Denial of service (DoS): Ping of Death and SYN Flood are examples of this kind of UIT; 2) Remote to local (R2L): SQL injection is an example of this kind of UIT, where typical vulnerabilities that are exploited is buffer overflow and pure environment sanitation; 3) User to root (U2R): SQL injection is also example of this kind of UIT; 4) Probe (Scanning): Nmap, IPswep, Satan are examples of software for scanning. During eight weeks, we simulate usual network traffic and UIT among hosts in each site. Normal network traffic and UIT were also simulated among sites. H-IDPS (NIST SP800-94, 2007) was installed in each one of the hosts. N-IDPS (NIST SP800-94, 2007) was installed at the gateway. Fig. 8 illustrates the sites, hosts with normal activities and infected hosts. Infected hosts inflict UIT to the hosts of each site and to hosts from other sites, as pointed by arrows. In this prototype, the propagation of UIT was in the following sequence: from site A to site A', from site A and A' to site A'', from site A, A' and A'' to site A. For this prototype, IFS was developed in JAVA and it runs in the three levels of DIFS. The IDPS Snort was used to analyze the network traffic. All classified UIT is lately recorded in a MySQL database. IFS collects data from the database, analyzes them and next, when a particular threshold of UIT is exceeded, a warning is sent to the IFS collaborators.

For the second prototype (Pontes et al, 2009), the two stage system was implemented and employed in a wired LAN, specifically in a computer working as gateway for the Internet (level 3 of the DIFS). Elements of level 1 (logs from the OS) were used in the. Although level 3 of the DIFS was approached, level 1, 2 and 4 were disregarded in the second prototype. The reason for implementing only level 3 is the representativeness of the gateway level: (a) the simulated cyber-attacks and the real network traffic have just one path to reach the Internet: throughout the gateway; (b) at the gateway level it was possible to assure timestamp conditions for correlation processes, as the IDPS is set at the same machine, the EAS and the gateway.

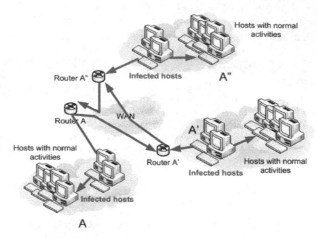

Fig. 8. DIFS Prototype – adapted from (Pontes & Guelfi, 2009)

Fig 9 illustrates the LAN for the tests, which is based on the diversity: diverse machines, settings, protocols and services are executed; further more there are several OS and free access to the Internet. Virtualized OSs (Linux Fedora), using VMWare, the host operational systems with Windows 7 and Windows XP are used in the prototype.

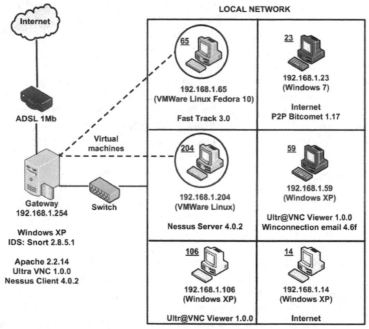

Fig. 9. Environment for Tests of the TSS – Computers, OSs and Services (Pontes et al, 2011)

The computer working as gateway (DIFS level 3) was able to register all alerts of the Network IDPS and logs from its own OS. Table 2 details services used in the two stage system, as the source machines for each service and the reached destiny for each service. In the environment for the tests, multi-correlation was done between alerts from an IDPS with the OS' logs.

Services	Source machine	Destiny
Internet browser	14 and 23	Internet
Remote access (VNC)	59 and 106	Gateway
Peer-to-peer (Bitcomet)	59	Internet
E-mail server (Winconnection)	59	Internet
Complete attack test (Fast-Track)	65	Gateway
Complete attack test (Nessus)	204	Gateway

Table 2. Services in the Prototype (Pontes et Al, 2011)

Table 3 presents applications which were used in the prototype. EAS was developed by the authors, in Visual FoxPro. Finally, Table 3 shows the elapsed time for the prototype. Both simulation of normal network traffic and simulation of cyber-attacks were referred in the prototype. Normal network traffic was brought up as well. Unlike (Pontes et al, 2008), (Pontes & Guelfi, 2009a), (Pontes & Guelfi, 2009b), (Pontes & Zucchi, 2010) Cyber-attacks concern the following types: (1) AWStats - allows remote attackers to execute arbitrary commands via shell; (2) SNMP: remote attackers can cause a DoS or gain privileges via SNMPv1 trap handling (SNMP AGENTX/TCP REQUEST is an example of this kind of attack); (3) P2P: multiple TCP/IP and ICMP implementations allow remote attackers to cause a DoS (reset TCP connections) via spoofed ICMP error messages.

Features	Applications	Time (m)	Details
EAS	Visual FoxPro		
IDPS	Snort	19	13113 signatures
logs Detection	Procmon	19	752851 logs
Graphs	Graphviz		

Table 3. Experiment Applications (Pontes et Al, 2011)

The following hardware were used for our prototype: gateway - Intel Core 2 Duo 2.66 GHz, 3 GB RAM – Windows XP Professional; number 65 – VMWare Workstation 6.0.2 768 MB RAM – Fedora 10 (Fast Track); number 204 - VMWare Workstation 6.0.2 512 MB RAM – Fedora 10 (Nessus Server 4.0.2); number 14 – Intel Core 2 Duo 2.5 GHz, 4 GB RAM – Windows XP Professional (browser's Internet access); number 23 – Intel Pentium 4 3.2 GHz, 4 GB RAM – Microsoft Windows 7 (browser's Internet accesss, Bitcomet 1.17); number 59 – Intel Pentium 4 3.06 GHz, 1 GB RAM – Windows XP Professional (Winconnection E-mail Server 4.6f, Ultr@VNC Viewer 1.0); number 106 – Intel Pentium 4 2,4 GHz, 2 GB –Windows XP Prof. (Ultr@VNC Server 1.0).

It is important to notice that the cyber-attacks considered in this prototype are, in matter of fact, a set of events (alerts and logs) classified as a single and more elaborated attack. In our earlier works (Pontes et al, 2008), (Pontes & Guelfi, 2009a), (Pontes & Guelfi, 2009b), (Pontes & Zucchi, 2010), forecasting techniques considered just individual events in the cyber-security context. Consequently in this paper forecasting techniques are differently employed, considering the DIFS architecture, as the prototype deals with more refined sets of attacks. Details regarding the EAS and the IFS tasks are not reported in this chapter due space limitations, but the reader may consult (Silva & Guelfi, 2010), (Silva, 2010) and (Pontes et al, 2008), (Pontes & Guelfi, 2009a), (Pontes & Guelfi, 2009b), (Pontes & Zucchi, 2010) for more information relating to EAS and IFS, respectively.

5. Results

Table 4 depicts the results of forecasting UIT in the first prototype. The UIT hit 4.320 thresholds from site A to site A' and, gradually, it increased with propagation of the UIT among the three sites. The total amount of the UIT thresholds among the three sites was about 16.416. In Table 4, correct forecasts are the number of times that it was possible to foresee the increasing and/or decreasing UIT's phases, without any delay. The correct predictions' rates were about 60,71%. Forecast with delay are the number of the times the increasing and decreasing thresholds were identified lately. In this prototype, forecasts' rates with delay were about 34,74%. During the prototype tests, sometimes it was not possible to identify thresholds for of UIT decreasing or increasing. The rate for the times we could not predict was about 4,95%.

	A →A'	A →A' →A"	A →A' →A" →A
Overall UIT thresholds	4.320	8.208	16.416
Correct forecast	2.623	4.984	9.967
Forecast with delay	1.483	2.818	5.635
Times not predict	214	406	814

Table 4. Results of Forecasting the UIT Propagation Using EWMA and Fibonacci Sequence (Pontes et al, 2009)

Table 5 depicts the results of forecasting UIT with only one forecasting technique (Fibonacci sequence) to the same experiments. The correct predictions' rates were among 5,21% and 7,55%. Forecasts with delay were among 3,92% and 4,68%.

	A →A'	A →A' →A"	A →A' →A" →A
Overall UIT thresholds	4.320	8.208	16.416
Correct forecast	326	534	855
Forecast with delay	195	384	643
Times not predict	3.799	7.290	14.918

Table 5. Results of Forecasting the UIT Propagation Using Only Fibonacci Sequence (Pontes et al, 2009)

In the second prototype, for the first step (EAS), results are achieved by analyzing consecutive graphs and tables from each phase. Quantity of alerts and correlations are independently accounted, according to the registered route (source and destination). In case the alerts and correlation regards the gateway, whether for source or destination), they are registered as Gateway; the alerts and correlation which disregard the gateway are registered as Non-Gateway. Table 6 summarizes the prototype and some results. Correlation shows a range of attack strategies. In each strategy a number of different alerts are connected sequentially as they were a single attack. A peer-to-peer (P2P) attack performed on machine 23 was chosen for the analysis of forecasting (Fig. 10, Fig. 11, Fig. 12 and Fig. 13).

	Values		Total
Detected alerts	2554 Gateway	1588 Non-gateway	4142
Alert types			137
Isolated alerts	29 (all in FP1)	21 FP / 8 TP	72,41% FP
Correlated alerts	14 FP1 = 21.08%	55 FP3 = 44.72%	
% FP	21.08 (FP1)	44.72% (FP3)	54.22%
% TP	45.78%	10.52% of all TP alerts were isolated	

Table 6. Prototype Results (Pontes et al, 2009)

Fig. 10 depicts the amount of FP which was detected, considering a preliminary correlation without FP filters. Notice there are 17 alerts (nodes) with 69 correlations among them (connections between alerts represented by arrows). Fig. 10 denotes the first scenario for comparisons: the DIFS level 3 work\ing without EAS.

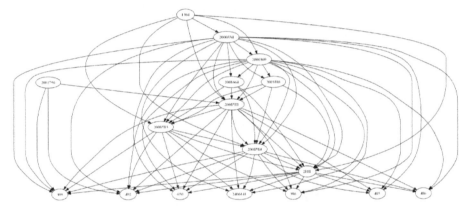

Fig. 10. P2P Graph Attack (TP + FP alerts) (Pontes et al, 2009)

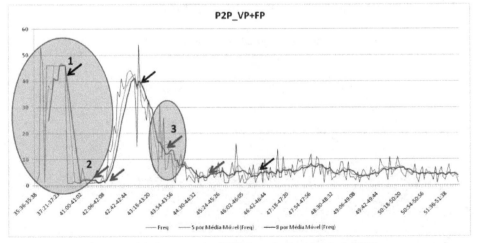

Fig. 11. True Positives + False Positves for P2P Attack (Pontes et al, 2009)

Fig. 11 illustrates the forecasting for cyber-attacks before the use the EAS, specifically for P2P events. Thus Fig. 11 takes into account the same scenario of Fig. 10. The ellipse spots the high volume of FP at the beginning of the experience with the prototype, consequently it is possible to notice three false thresholds for the forecasting, as shown by points (1), (2) and (3). Forecasting was done by the use of diverse EWMA.

Fig. 12 represents the graph after applying EAS filtering. Notice there are just 8 alerts (nodes) with 22 correlations among them (connections between alerts represented by arrows). Fig. 12 denotes the second scenario for comparisons: the DIFS level 3 (gateway level) working with the EAS filtering. As a result by the use of EAS, it was possible to track FP, filtering them, in order to improve forecasts, as the false thresholds for the predictions were eliminated as well.

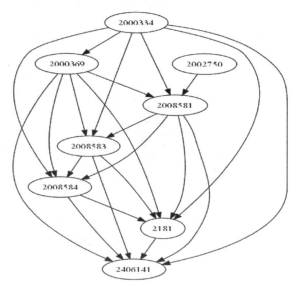

Fig. 12. P2P Graph Attack (only TP alerts) (Pontes et al, 2009)

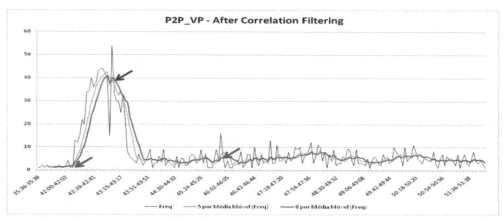

Fig. 13. True Positives for the P2P Attack – After the Correlation Filtering (Pontes et al, 2009)

Fig. 13, in the next page, depicts the application of forecasting techniques (diverse EWMA), i.e. the IFS, after the employment of EAS filtering. In Fig. 13 it is possible to verify two thresholds pointing out the increasing of events (as indicated by the red arrows, and one threshold point out the decreasing of events (as shown by black arrow). Notice there is no significant occurrence of alerts at the beginning of the experiment and two false thresholds regarding forecasts were eliminated. It is also important to observe that the second ellipse with the FP were eliminated after the EAS filtering, hence, another false threshold was wipe out as consequence. More details regarding results can be found in (Pontes et al 2011).

6. Conclusion

As a conclusion, this chapter has introduced the Distributed Intrusion Forecasting System (DIFS) (Pontes et al, 2009), approaching cyber attacks and UIT in the cyber space context. The DIFS also presented the two stage system with the EAS implemented for making the multi-correlation (step 1) (Pontes et al, 2009), afterwards the application of the forecasting techniques over the generated data by the EAS (step 2). The forecasting model presented in this chapter could be analogously employed for earthquake prediction, due the following aspects: a) DIFS, with the Two Stage System and the EAS, was able to track in advance the increasing and decreasing rates of cyber attacks and UIT; hence such methodology may be employed as an early warning system; b) DIFS considers just frequency and temporal characteristics (timestamp) of events (UIT and cyber attacks), thus this approach can be similarly used in other areas.

Even though only 4,95% of the thresholds for UIT's increasing and decreasing were not detectable, the value of the outcome is still questionable, as this early warning system still has 34,74% of warnings being lately reported. The use of two forecasting techniques represented better results if compared to the use of only one prediction technique. The reason for the accuracy using two forecasting techniques, according to (Pretcher and Frost, 2002), is due to the fact Fibonacci sequence depends on EWMA for marking the first wave. Thus, it was possible to observe just some of the trends drew by the Fibonacci sequence. Another characteristic for predictions with Fibonacci sequence is forecasts in the long term (2, 3 days): EWMAs don't have this feature, so, predictions using only EWMA lack in long term predictions. Employing both of the techniques aggregates the positive of either techniques, making the forecast more accurate.

For the EAS, it was suggested a standard to define causes and consequences within the PC-correlation method combined with multi-correlation criteria, correlation analysis (ascending/descending) and identification of FP alerts through tables and graphs. It was done an experiment with a prototype, in a LAN, with diverse machines and OS, which used a gateway to get access to the Internet. The obtained results from the tests in our prototype indicate that level 3 of DIFS was improved, as some FPs were treated and predictions concerning cyber-attacks were more accurate. It is possible to come to this conclusion by verifying that, despite high FP rates of FP1 (21.08%) and FP3 (44.72%) – see Table III -; during the whole experiment, no TP alert was correlated exclusively as result of an FP alert.

As a suggestion for improving the work, it is suggested to automate analysis' processes that require user interpretation (table correlation and mapping) for using the EAS in real time.

The accuracy of the results can be improved whether the multi-correlation is extended to entire LAN. Regarding the forecast's result, among the suggestions for future works there are the aggregation of the fractal approaches (according to (Mandelbrot & Hudson, 2006)), and the use of other kinds of forecasting techniques (as Markov chains and neural networks) to follow (Armstrong, 2002)'s advices. It is also suggested to extend the employment of the EAS for the four levels of DIFS, so levels 1, 2 and 4 may be approached in future works. The EAS/DIFS has not yet undergone extensive training enough to be used in commercial applications.

7. References

A. A. A. Silva, "A security event analisys system to identify false positive alerts and evaluate isolated alerts creating multi-correlation criteria". IPT; São Paulo, SP, Brasil, 2010, 107 f. Masters Dissertation in Computer Engeneering.

Abad, Cristina et al. "Log correlation for intrusion detection a proof of concept".. p. 10. In the 19th IEEE ACSAC 2003. University of Illinois as Urbana-Champaign; 2003, ISBN 0-7695-2041-3, pp. 8-12.

Andersson, L.; Davies, E.; Zhang, L.; 2007 "Report from the IAB workshop on Unwanted Traffic March 9-10, 2006", RFC 4948, IETF. 2006

Alampalayam, P.; Kumar, A. "predictive security model using data mining", in proc IEEE Globcom, 2004.

Armstrong, J. S.. Principles of Forecasting: A Handbook for Researchers and Practitioners, (2002). Springer (Ed), ISBN 0792379306, USA, 2002

Bleier, T.; Freund, F. (2005). Earthquake [earthquake warning systems], In IEEE Spectrum. Vol 42, Issue 12, (05 December 2005) , pp. 22, ISSN 0018-9235.

Chung, Y.; Kim, I.; Lee, C.; Im, E. G.; Won, D. "Design of on-line intrusion forecast system with a weather forecasting model", In the Springer ICCSA 2006.

Cisar, P.; Cisar, S. M. "EWMA Statistic in Adaptive Threshold Algorithm", In the IEEE INES, 2007, pp 51-54.

Feitosa, E. L.; Souto, E. J.; Sadok, D. "Tráfego Internet não Desejado: Conceitos, Caracterização e Soluções". in Proc. VIII SBSeg, SBC. 2008 pp. 91-137.

Gula, Ron. Correlating IDS Alerts with Vulnerability Information. Chief Technology Officer – Tenable Network Security; Columbia, MD, EUA, 2007. p. 10.

Haslum, K.; Abraham, A.; Knapskog, S., (2008). Fuzzy Online Risk Assessment for Distributed Intrusion Prediction and Prevention Systems, Proceedings of IEEE UKSIM 2008 10th International Conference on Computer Modeling and Simulation, pp. 216-223, ISBN 0-7695-3114-8, Cambridge, UK, April 1-3, 2008

Holliday, James R.; Nanjo, Kazuyoshi Z.; Tiampo, Kristy F.; Rundle, John B.; Turcotte, Donald L.; (2005). Earthquake forecasting and its verification, Nonlinear Processes in Geophysics,.

IC3 - Internet Crime Complaint Center, "2009 Internet Crime Report" Bureau of Justice Assistance and National White Collar Crime Center, 2010, [Online]. Available: www.ic3.gov, 2010.

Ishida, C.; Arakawa, Y.; Sasase, I. "Forecast Techniques for Predicting Increase or Decrease of Attacks Using Bayesian Inference", In the IEEE PACRIM, 2005, pp 450-453.

Jemilli, F., Zaghdoud, M.; Ahmed, M. B. "DIDFAST.BN :Distibuted intrusion detection and forecasting multiagent system using bayesian network", In the IEEE ICTTA, 2006, pp 3040-3044.

King, Samuel; Chen, Peter; Mao, Z. Morley; Lucchetti, Dominic G. "Enriching intrusion alerts through multi-most causality". In the 12th NDSS 2005, University of Michigan, USA, 2005. p. 13.

Lai-Chenq, C. "A high-efficiency intrusion prediction technology based on markov chain", In the IEEE CISW, 2007.

Lajara, R., Alberola, J., Pelegri, J., Sogorb, (2007) Ultra Low Power Wireless Weather Station, Proceedings of IEEE SENSOR COMM International Conference on Sensor

Technologies and Applications, pp. 469-474, ISBN 978-0-7695-2988-2, Valencia, Spain, October 14-20, 2007

Leu, F.; Yang, W.; Chang, W. "IFTS : Intrusion Forecast and Traceback based on Union Defense Environment", In the IEEE ICPADS, 2005.

Lorenz, E. N. "Designing chaotic models", Journal of the Atmospheric Sciences: Vol. 62, No. 5, ISSN 1520-0469, 2005, pp. 1574–1587.

Mandelbrot, B.; Hudson, R. L. "The behavior of markets: a fractal view of risk, ruin and reward", John Willey, 2006.

Manikopoulos, Constantine; Papavassiliou; Symeon, Network Intrusion and Fault Detection: A Statistical Anomaly Approach. In IEEE Communications Magazine 40, 2002, pp. 76-82 New Jersey Institute of Technology, NJ, EUA, 2002. p. 7.

McPherson, D.; Labovitz, C. "5th Worldwide Infrastructure Sec. Report", 2010, [Online]. Available: http://seclists.org/funsec/2010/q1/295 /, 2010.

Mizoguchi, Fumio, Anomaly Detection using Visualization and Machine Learning. In the IEEE 9th International WET ICE, 2000, pp. 76-82. Science University of Tokyo – Information Media Center; Noda, Japan, 2000. p. 6.

Morin, Benjamin; Debar, Hervé. Correlation of Intrusion Symptoms: An Application of Chronicles. France Télécom R&D; In the 6th International Conference on RAID, 2003, PP. 94-112. Springer-Verlag - Berlin Heidelberg , 2003, G. Vigna, E. Jonsson, and C. Kruegel (Eds.).

Ning, Peng; Cui, Yun. "An intrusion alert correlator based on prerequisites of intrusions". Technical Report TR-2002-01 North Carolina State University; Raleigh, NC, USA, 2002. p. 16.

Ning, Peng; Cui, Yun; Reeves S., Douglas; Analyzing Intensive Intrusion Alerts via Correlation. North Carolina State University; In the 5th International Symposium on RAID, 2002, p. 21. Raleigh, NC, EUA, .

NIST – National Institute of Standards and Technology, (2007). Guide to Intrusion Detection and Prevention Systems (IDPS), In: NIST SP 800-94, December, 2010, Available from: <http://csrc.nist.gov/publications/>

NIST/SEMATECH, e-Handbook of Statistical Methods, 2009, www.itl.nist.gov/.

Pietraszek, Tadeusz; Tanner, Axel. Data mining and Machine Learning – Towards Reducing False Positives in Intrusion Detection. IBM Zuurich Research Laboratory, Ruschlikon, Suécia, 2005. Information Security Technical Report, Vol. 10, ed. 3, pp 169-183.

Pontes, E.; Guelfi, A. E., "Third generation for intrusion detection: applying forecasts and ROSI to cope with unwanted traffic". In Proceedings of 4th IEEE ICITST 09, London, UK, November 2009, ISBN 978-1-4244-5647-5, pp. 1-6.

Pontes, E.; Guelfi, A. E.; Alonso, E. "Forecasting for return on security information investment: new approach on trends in intrusion detection and unwanted traffic". In IEEE Journal Latin America Transactions, 2009, Vol 7, ISSN 1548-0992, pp 438-445.

Pontes, E.; Guelfi, A., (2009). IFS – Intrusion forecasting system based on collaborative architecture, Proceedings of the IEEE ICDIM 2009 4th International Conference on Digital Information Management, pp. 1-4, ISBN 978-1-4244-4253-9, Ann Arbor, Michigan, USA, Nov 1-4, 2009

Pontes, E.; Zucchi, W. L. "Fibonacci sequence and EWMA for intrusion forecasting system". In 5th ICDIM 2010, Lakehead University, Thunder Bat, Canada, July 2010, ISBN 978-1-4244-7571-1, pp. 1-6.

Pontes, E.; Guelfi, A. E., Silva, A. A. A., Kofuji, S. T. "Applying Multi-Correlation for Improving Forecasting in Cyber Security". In 6th ICDIM 2011, Melbourne University, Thunder Bat, Canada, July 2010, ISBN 978-1-4244-7571-1, pp. 1-6.

Prechter, R. R. Jr; Frost, A. J "Elliott Wave Principles", John Wiley, 2002.

Ramasubramanian, P.; Kannan, A. "Quickprop neural network ensemble forecasting framework for database intrusion prediction system", In the Springer 7th ICAISC, 2004, pp 9-18.

Reeves S., Douglas; Ning, Peng; Cui, Yun. "Constructing attack scenarios through correlation of intrusion alerts". In the 9th ACM CCCS, North Carolina State University; CCS'02, Washington, DC, USA., 2002. p. 10.

Roberts, S. W. "Control Chart Tests Based On Geometric Moving Average", Technometrics, pages 239-251,1959.

Silva, A. A. A.; "A security event analisys system to identify false positive alerts and evaluate isolated alerts creating multi-correlation criteria". IPT; São Paulo, SP, Brasil, 2010, 107 f. Masters Dissertation, Computer Engineering Department.

Silva, A. A. A.; Guelfi, A. E. "Sistema para identificação de alertas falso positivos por meio de análise de correlacionamentos e alertas isolados", In the 9th IEEE I2TS 2010, Rio de Janeiro, Brazil, 2010.

Sindhu, S.S.S.; Geetha, S.; Sivanath, S.S.; Kannan, A. "A neuro-genetic ensemble short term forecasting framework for anomaly intrusion prediction", In the IEEE ADCOM, 2006, pp 187-190.

Su, You-Po; Zhu, Qing-Jie (2009). Application of ANN to Prediction of Earthquake Influence, Proceedings of IEEE ICIC '09 Second International Conference on Information and Computing Science, pp. 234 – 237, ISBN 978-0-7695-3634-7, Manchester, UK, May 21-22, 2009

Valdes, Alfonso; Skinner, Keith; Probabilistic Alert Correlation. SRI International; In the 2001 International Workshop on the RAID, 2001, pp. 54-68. Springer-Verlag Berlin Heidelberg , 2001, W. Lee, L. Me, and A. Wespi (Eds).

Viinikka, J.; Debar, H.; Mé, L.; Séguier, R. "Time Series Modeling for IDS Alert Management", A In the CM ASIAN ACM Symposium on Information, Computer and Communications Security, 2006.

Wong, W.; Guan, X., Zhang, X.; Yang, L. "Profiling program behavior for anomaly intrusion detection based on the transition and frequency property of computer audit data", In the ELSEVIER Computer & Security, 2006.

Ye, N.; Li, X.; Chen, Q.; Emran, S. M.; Xu, M. "Probabilistic techniques for IDS based on computer audit data", In the IEEE Transactions on Systems, Man and Cybernetics, pages 266-274, IEEE, 2001.

Ye, N.; Vilbert, S.; Chen, Q. "Computer intrusion detection through EWMA for autocorrelated and uncorrelated data", In the IEEE Transactions on Reliability, pages 75-82, IEEE, 2003.

Yin, Q.; Shen, L.; Zhang, R.; Li, X "A new intrusion detection method based on behavioral model", In the IEEE WCICA, 2004, pp 4370-4374.

Zhay, Yan; Ning, Peng; Iyer, Purush; Reeves, Douglas S. "Reasoning about complementary intrusion evidence", In 20th Annual CSAC. North Carolina State University;USA, 2004. pp. 39-48.

Zhay, Yan; Ning, Peng; Xu, Jun "Integrating IDS alert correlation and os-level dependency tracking". Technical Report TR-2005-27 North Carolina State University , 2006, S. Mehrotra et al. (Eds.): ISI 2006, LNCS 3975, pp. 272–284, 2006.

Zhengdao, Z.; Zhumiao, P.; Zhiping, Z. "The Study of Intrusion Prediction Based on HSMM", In the IEEE Asia-Pacific Services Computing Conference, 2008, pp 1358-1363.

Zuckerman, E.; Roberts, H.; McGrady, R.; York, J.; Palfrey, J. "distributed denial of service attacks against independent media and human rights sites", 2010, [Online]. Available: http://www.soros.org, 2010.

7

Application of Recurrent Radon Precursors for Forecasting Local Large and Moderate Earthquakes

Ming-Ching T. Kuo
National Cheng Kung University
Taiwan

1. Introduction

Measurement of radon-222 in groundwater has been frequently used in earthquake prediction (Igarashi et al. 1995; Liu et al. 1985; Noguchi & Wakita 1977; Teng, 1980; Wakita et al. 1980; Kuo et al. 2006, 2010a, 2010b). According to a worldwide survey (Hauksson 1981; Toutain & Baubron 1999), more than 80 % of radon (Rn-222) anomalies associated with earthquakes show increases in radon concentration precursor to a rupture while a few anomalies manifested decreases in radon. The purpose of this chapter is to provide a practical guide of monitoring groundwater radon for the early warning of local disastrous earthquakes. In this chapter, methods of monitoring groundwater radon including procedures of sample collection and radon determination will be addressed. The following sections outline suitable geological conditions to consistently catch precursory declines in groundwater radon, in-situ radon volatilization mechanism for interpreting anomalous decreases in groundwater radon prior to earthquakes, and mathematical model for quantifying gas saturation developed in newly created cracks preceding an earthquake. Case studies are provided to illustrate the application of recurrent radon precursors for forecasting local large and moderate earthquakes.

2. Sample collection radon determination

Accurate sampling for radon measurements depends on appropriate monitoring wells. Because radon concentration in groundwater relates to emanation rates of geological layers, representative sampling must be from properly constructed wells. A submersible pump is commonly used in monitoring wells for groundwater sampling except artesian wells. Every sampling starts with flushing the stagnant water in the well and especially in the screen zone. Inadequate purging can be a major source of error, because the water sample is a mixture of stagnant water from the well bore, pore water from the filter gravel and groundwater influenced by the natural emanation rate of the aquifer. Fig. 1 shows the radon concentration in the well discharge during continuous sampling in a monitoring well. During the first period of flushing, the radon concentration of the water samples is practically zero and then increases rapidly to 529 pCi/L. The mean radon concentration measured for this monitoring well was 529 ± 19 pCi/L (eleven samples). A minimum of 3 well-bore volumes was purged before taking samples for radon measurements.

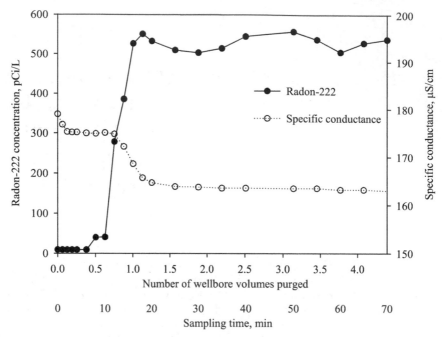

Fig. 1. Radon concentration and electrical conductivity in the well discharge during continuous sampling in a deep observation well

A 40-ml glass vial with a TEFLON lined cap was used for sample collection. After collecting a sample, the sample vial was inverted to check for air bubbles. If any bubbles were present, the sample was discarded and the sampling procedure repeated. The date and time of sampling was recorded and the sample stored in a cooler. The maximum holding time before analysis was 4 days.

For the determination of the activity concentration of radon-222 in groundwater, a modified method described by Prichard and Gesell (1977) was adopted. Radon was partitioned selectively into a mineral-oil scintillation cocktail immiscible with the water sample (Noguchi 1964). The sample was dark-adapted and equilibrated, and then counted in a liquid scintillation counter (LSC) using a region or window of the energy spectrum optimal for radon alpha particles (Lowry 1991).

Radon concentrations were determined by drawing a 15-ml sample directly from a field sample into a clean syringe. Care was taken to prevent aeration of the samples in the process. The samples were then injected beneath a 5-ml layer of mineral-oil-based scintillation solution in 24-ml vials. The vials were vigorously shaken to promote phase contact, dark-adapted and held for at least three hours to ensure equilibrium between radon-222 and its daughters, and then assayed with a liquid scintillation counter. The results were corrected for the amount of radon decay between sampling and assay.

The results of the measurements were determined in units of counts per minute (cpm). It was essential to ensure that only the activity of radon-222 was measured. Using the TRI-CARB software of Packard 1600TR, it was possible to view the alpha spectrum (Fig. 2). The peaks of radon-222 (5.49 MeV), polonium-218 (6.00 MeV) and polonium-214 (7.69 MeV) can be distinguished.

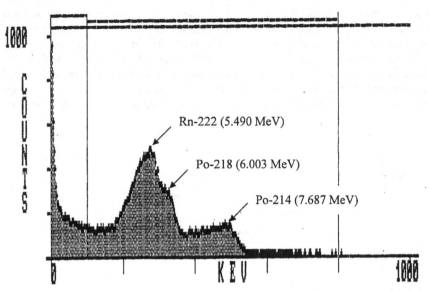

Fig. 2. Alpha spectrum of radon-222 and its daughter nuclides represented by TRI-CARB software

A calibration factor for the LSC measurements of 7.1 ± 0.1 cpm/pCi (Fig. 3) was calculated using an aqueous Ra-226 calibration solution, which is in secular equilibrium with Rn-222 progeny. For a count time of 50 min and background less than 6 cpm, a detection limit below 18 pCi/L was achieved using the sample volume of 15-ml (Prichard et al. 1992).

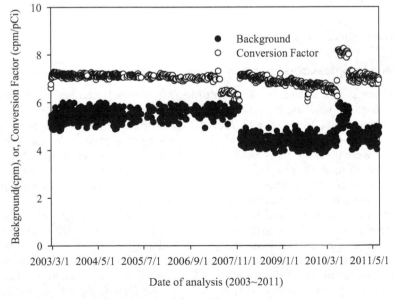

Fig. 3. Calibration factor and background for LSC measurements

Verification of radon-222 as the radioisotope responsible for activity in the well water tested was obtained by the repeated counting of three samples from two wells. The half-life of 3.841 days experimentally determined for samples from Well Liu-Ying (I) located in Tainan Plain, Taiwan compares favorably with the accepted value of 3.825 days as shown in Fig. 4. When the counting vials are lack of tightness, radon will escape from counting vials and the half-life times experimentally determined for samples will be apparently shorter. Fig. 4 also shows an example of such a case from Well Wen-Tsu (II) located in Choshui River Alluvial Fan, Taiwan.

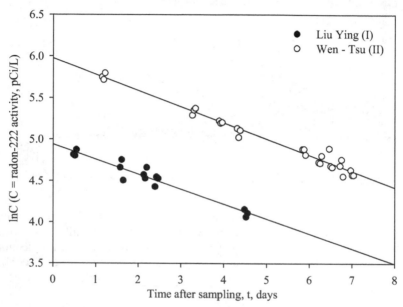

Fig. 4. Measurement of half life from semi-logarithmic decay curve

3. Suitable geological conditions to catch recurrent radon precursors

The 2003 Chengkung earthquake of magnitude (M) 6.8 on December 10, 2003 was the strongest earthquake near the Chengkung area in eastern Taiwan since 1951. The Antung radon-monitoring well (D1, Fig. 5) was located 20 km from the epicenter. Approximately 65 days prior to the 2003 Chengkung earthquake, precursory changes in radon concentration in ground water were observed. Specifically, radon decreased from a background level of 780 pCi/L to a minimum of 330 pCi/L (Fig. 6). Both geological conditions near the Antung hot spring and the vapor-liquid phase behavior of radon were investigated to explain the anomalous decrease of radon precursory to the 2003 Chengkung earthquake.

The production interval of the well ranges from 167 m to 187 m below ground surface and is pumped more or less continuously for water supply purposes. Discrete samples of geothermal water were collected for analysis of radon (Rn-222) twice per week. Liquid scintillation method was used to determine the activity concentration of radon-222 in ground water (Noguchi 1964; Prichard et al. 1992). The radon concentration was fairly stable (780 pCi/L in average) from July 2003 to September 2003 (Fig. 6). Sixty-five days before the magnitude (M) 6.8 earthquake (December 10, 2003), the radon concentration of ground water started to decrease and continued to decrease for 45 days. Twenty days prior to the earthquake, the radon concentration reached a minimum value of 330 pCi/L and before starting to increase. Just before the earthquake, the radon concentration recovered to the previous background level of 780 pCi/L. The main shock also produced a sharp anomalous coseismic decrease (~300 pCi/L). After the earthquake, some irregular variations were observed, which we interpret as an indication that the strain release by the main shock was not complete and that some accumulation and release of strain continued in the region.

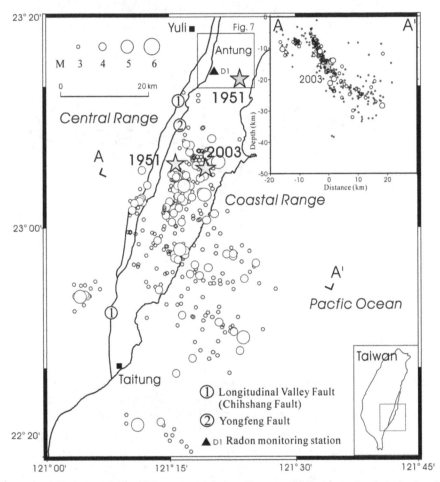

Fig. 5. Map of the epicentral and hypocentral distributions of the mainshock and aftershocks of the 2003 Chengkung earthquake and 1951 mainshocks (star: mainshock, open circles: aftershocks).

The Antung hot spring (Fig. 7) is in a unique tectonic setting located at the boundary between the Eurasian and Philippine Sea plates near the Coastal Range. Four stratigraphic units are present. The Tuluanshan Formation consists of volcanic units such as lava and volcanic breccia as well as tuffaceous sandstone. The Fanshuliao and Paliwan Formations consist of rhythmic sandstone and mudstone turbidites. The Lichi mélange occurs as a highly deformed mudstone that is characterized by penetrative foliation visible in outcrop. The Antung hot spring is situated in a brittle tuffaceous-sandstone block surrounded by a ductile mudstone of the Paliwan Formation (Chen & Wang 1996). Well-developed minor faults and joints are common in the tuffaceous-sandstone block displaying intensively brittle deformation. It is possible that these fractures reflect deformation and disruption by the nearby faults. Hence, geological evidence suggests the tuffaceous-sandstone block displays intensively brittle deformation and develops in a ductile-deformed mudstone strata. Ground water flows through the fault zone and is then diffused into the block along the minor fractures.

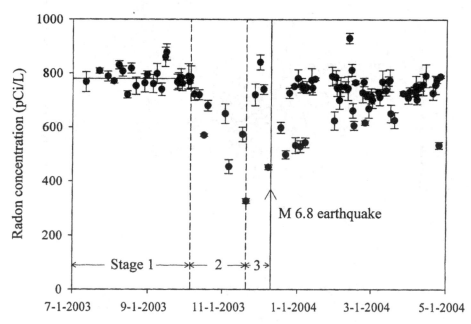

Fig. 6. Radon concentration data at the monitoring well (D1) in the Antung hot spring. Stage 1 is buildup of elastic strain. Stage 2 is dilatancy and development of cracks and gas saturation. Stage 3 is influx of ground water and diminishment of gas saturation.

Under geological conditions such as those of the Antung hot spring, we hypothesized that when regional stress increases, dilation of the rock mass occurs at a rate faster than the rate at which pore water can flow into the newly created pore volume (Brace et al. 1966; Scholz et al. 1973). During this stage (Stage 2 in Fig. 6), gas saturation and two phases (vapor and liquid) develop in the rock cracks. Meanwhile, the radon in ground water volatilizes and partitions into the gas phase and the concentration of radon in ground water decreases. Thus, the sequence of events for radon data prior to the 2003 Chengkung earthquake (Fig. 6) can be interpreted in three stages. From July 2003 to September 2003 (Stage 1), radon was fairly stable (around 780 pCi/L). During this time, there was an accumulation of tectonic strain, which produced a slow, steady increase of effective stress. Sixty-five days before the magnitude (M) 6.8 earthquake, the concentration of radon started to decrease and reached a

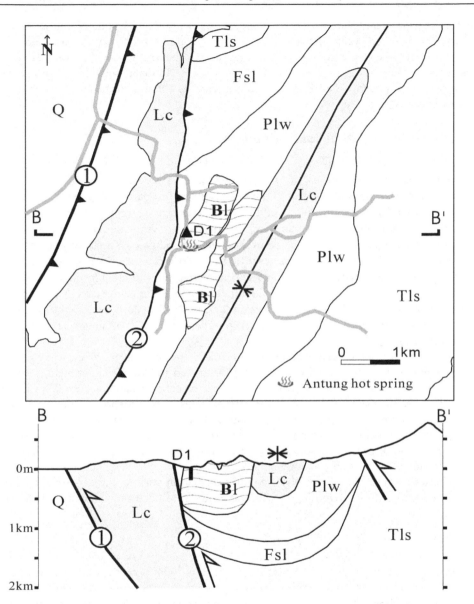

Fig. 7. Geological map and cross section near the radon-monitoring well in the area of Antung hot spring (Q: Holocene deposits, Lc: Lichi mélange, Plw: Paliwan Formation, Fsl: Fanshuliao Formation, Tls: Tuluanshan Formation, Bl: tuffaceous fault block, D1: radon-monitoring well, : Chihshang, or, Longitudinal Valley Fault, : Yongfeng Fault). See Fig. 6 for map location.

minimum value of 330 pCi/L twenty days before the earthquake. During this 45-day period (Stage 2), dilation of the rock mass occurred and gas saturation developed in cracks in the rock and radon volatilized into the gas phase. Stage 3 started at the point of minimum radon concentration when water saturation in cracks and pores began to increase and radon increased and recovered to the background level. The main shock produced a sharp coseismic anomalous decrease (~300 pCi/L). After the earthquake, some irregular variations were observed, which we attribute to strain release as some accumulation and release of strain continued in the region.

4. In-situ radon volatilization mechanism

Radon partitioning into the gas phase can explain the anomalous decreases of radon precursory to the earthquakes (Kuo et al., 2006). To support the hypothesis of radon volatilization from ground water into the gas phase, radon-partitioning experiments were conducted to determine the variation of the radon concentration remaining in ground water with the gas saturation at formation temperature (60 ℃) using formation brine from the Antung hot spring. Five levels of gas saturation were investigated, specifically S_g = 5 %, 10 %, 15 %, 20 %, and 25 % where S_g is gas saturation. Triplicate experiments were conducted for each level of gas saturation. Every experiment started with 40-ml of formation brine. Five levels of headspace volume at 2 ml, 4 ml, 6 ml, 8 ml, and 10 ml were then created above the liquid phase for five levels of gas saturation at 5 %, 10 %, 15 %, 20 %, and 25 %, respectively. Two-phase equilibrium was achieved for each experiment in 30 minutes at the formation temperature (60 ℃).

A kinetic study of radon volatilization from ground water into the gas phase was conducted to determine the time required to reach equilibrium. In the kinetic experiment, formation brine from the Antung hot spring with an initial radon concentration of 479 ± 35 pCi/L was used. Every sample started with 40-ml formation brine and a headspace volume at 6 ml was then created above the liquid phase. A total of five samples were prepared. The radon concentration remaining in ground water was determined at various volatilization times (i.e., 2 min, 5 min, 15 min, 30 min, and 60 min). The time required to reach equilibrium for radon volatilization was only about 5 minutes.

Data from the vapor-liquid two-phase equilibrium radon-partitioning experiments (Fig. 8) were regressed with the two-phase partitioning model to determine Henry's coefficient as follows.

$$C_0 = C_w (H \times S_g + 1)$$ (1)

where C_0 is initial radon concentration in groundwater precursory to each radon anomaly, pCi/L; C_w is the radon minimum in groundwater observed in well D1 during an anomalous decline, pCi/L; S_g is gas saturation, fraction; H is Henry's coefficient for radon at formation temperature (60 ℃), dimensionless. Fig. 8 shows the regressed line with H = 12.8 and R^2 = 0.9919 (regression coefficient). Henry's coefficient for radon at 60 ℃ determined for the Antung formation brine (12.8) is higher than the value (7.91) for water at 60 ℃ (Clever, 1979). Fig. 8 can be used to estimate the amount of gas saturation required for various decreases in concentration of radon. For example, the anomalous decrease of radon concentration from 780 pCi/L to 330 pCi/L required a gas saturation of 10 % in cracks in the rock.

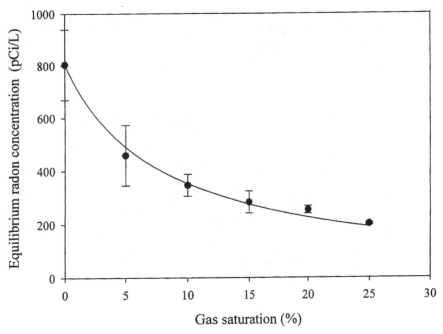

Fig. 8. Variation of radon concentration remaining in ground water with gas saturation at 60 °C using formation brine from the Antung hot spring.

5. Case study

We have monitored groundwater radon since July 2003 at well D1 at the Antung hot spring that is located about 3 km southeast of the Chihshang fault (Fig. 9). The Chihshang fault is part of the eastern boundary of the present-day plate suture between the Eurasia and the Philippine Sea plates. The Chihshang fault ruptured (Hsu, 1962) during two 1951 earthquakes of magnitudes M = 6.2 and M = 7.0. The annual survey of geodetic and GPS measurements has consistently revealed the active creep of the Chihshang fault that is moving at a rapid steady rate of about 2-3 cm/yr during the past 20 years (Angelier et al., 2000; Yu & Kuo, 2001; Lee et al., 2003).

Fig. 10 shows the radon concentration data since July 2003 at the monitoring well (D1) in the Antung hot spring. Radon-concentration errors are ±1 standard deviation after simple averaging of triplicates. Recurrent groundwater radon anomalies were observed to precede the earthquakes of magnitude M_w = 6.8, M_w = 6.1, M_w = 5.9, and M_w = 5.4 that occurred on December 10, 2003, April 1, 2006, April 15, 2006, and February 17, 2008 at the Antung D1 monitoring well. We consider the M_w 5.9 earthquake that occurred on April 15, 2006 triggered by stress transfer in response to the 2006 M_w 6.1 Taitung earthquake. All the three recurrent anomalous decreases observed at Antung follow the same v-shaped progression and are marked with green inverted triangles in Fig. 10. Environmental records such as atmospheric temperature, barometric pressure, and rainfall were examined to check whether the radon anomaly could be attributed to these environmental factors. The atmospheric temperature, barometric pressure, and rainfall are periodic in season. It is difficult to explain such a large radon decrease by these environmental factors. There was also no heavy rainfall responsible for the radon anomaly.

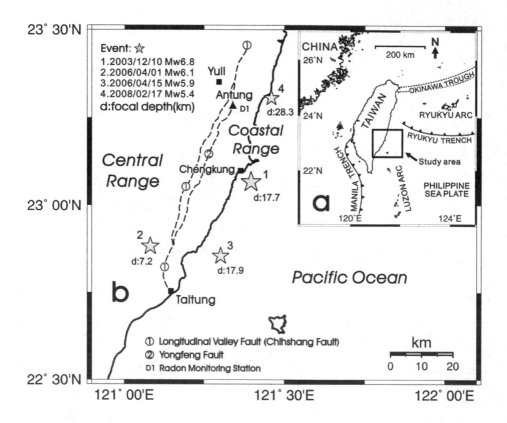

Fig. 9. Map of the epicenters of the earthquakes that occurred on December 10, 2003, April 1 and 15, 2006, February 17, 2008 near the Antung hot spring. (a) Geographical location of Taiwan. (b) Study area near the Antung hot spring.

The box-and-whisker plot is used on the right-hand side in Fig. 10. It shows the median (50th percentile, 764 pCi/L) as a center bar, and the quartiles (25th and 75th percentiles, 692 pCi/L and 849 pCi/L) as a box. The whiskers (456 pCi/L and 1077 pCi/L) cover all but the most extreme values in the data set. Based on the box-and-whisker plot, the threshold concentration of anomalous radon minima at the Antung D1 monitoring well is estimated as 456 pCi/L. The radon minimum recorded prior to the 2008 M_w = 5.4 Antung earthquake is close to the threshold concentration and can be easily masked by the noisy background. On the other hand, the radon anomalous minima, recorded precursory to strong earthquakes ($M_w > 6.0$), the 2003 M_w = 6.8 Chengkung and 2006 M_w = 6.1 Taitung earthquakes, are low enough to be clearly distinguished from the background noise.

The radon minima, measured prior to local moderate earthquakes, are easily masked by the noisy background. Fig. 10 also shows the large background variation in radon data following the 2003 M_w = 6.8 Chengkung, 2006 M_w = 6.1 Taitung, and 2008 M_w = 5.4 Antung earthquakes. Four local earthquakes with magnitudes (M_w) of 5.5, 5.2, 6.2, and 5.2 occurred on 12/11/2003, 1/1/2004, 5/19/2004, and 9/26/2005, respectively. Based upon their magnitudes and locations, we consider these as aftershocks and induced events of the 2003 Chengkung earthquake. The large scatter in radon data between the 2003 M_w = 6.8 Chengkung and 2006 M_w = 6.1 Taitung earthquakes can be related to these aftershocks. The 2006 Mw 6.1 Taitung earthquake also triggered the Mw 5.9 earthquake that occurred on April 15, 2006. One local earthquake of magnitude M_w = 4.9 that occurred on 6/4/2006 can be considered as an aftershock of the 2006 M_w = 6.1 Taitung earthquake. The M_w = 4.9 aftershock also caused a large scatter in radon data following the 2006 M_w = 6.1 Taitung earthquake. The large background variation in radon data following the 2008 M_w = 5.4 Antung earthquake can also be attributed to local earthquakes, such as a local earthquake of magnitude M_w = 5.3 that occurred on 12/2/2008.

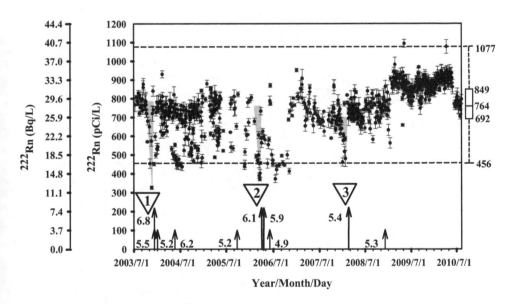

Fig. 10. Radon concentration data at well D1 in the Antung hot spring (open inverted triangles: anomalous radon minima; green inverted triangles: v-shaped pattern; long arrows: mainshocks; short arrows: aftershocks; earthquake magnitude M_w shown beside arrows).

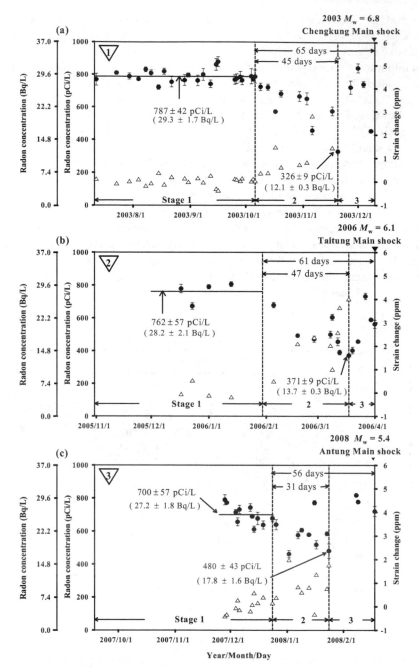

Fig. 11. Observed radon anomalies at well D1 prior to (a) 2003 Chengkung, (b) 2006 Taitung, and (c) 2008 Antung earthquakes. Stage 1 is buildup of elastic strain. Stage 2 is development of cracks. Stage 3 is influx of ground water.

The observed v-shaped pattern prior to the three main shocks clearly progresses in a sequence of three stages (Kuo et al. 2006). The sequence of events for radon anomalies prior to the 2003 M_w = 6.8 Chengkung, 2006 M_w = 6.1 Taitung, and 2008 M_w = 5.4 Antung earthquakes were characterized into three stages in Figs. 11a, 11b, and 11c, respectively (Kuo et al. 2006, 2010a). During Stage 1, the radon concentration in ground water was fairly stable; there was an accumulation of tectonic strain and a slow, steady increase of regional stress. The Antung hot spring is a fractured aquifer with limited recharge surrounded by ductile mudstone (Chen & Wang 1996). When the regional stress increased under these geological conditions, dilation of brittle rock masses occurred at a rate faster than the rate at which ground water could recharge into the newly created rock cracks (Brace et al. 1966; Nur 1972; Scholz et al. 1973). During this stage (Stage 2 in Fig. 11), gas saturation and two phases (vapor and liquid) developed in the rock cracks. The radon in ground water volatilized into the gas phase and the radon concentration in ground water decreased. Stage 3 started at the point of minimum radon concentration when the water saturation in cracks and pores began to increase again. During this stage (Stage 3 in Fig. 11), the radon concentration in groundwater increased and recovered to the previous background level before the main shock.

Figs. 11a, 11b, and 11c show that during Stage 2 prior to the 2003 M_w = 6.8 Chengkung, 2006 M_w = 6.1 Taitung, and 2008 M_w = 5.4 Antung earthquakes the radon concentration in ground water kept decreasing for a significantly long period of 45, 47, 31 days, respectively. Combining the use of box-and-whisker plot, the v-shaped radon pattern shown in Figs. 10 and 11 prior to the 2003 M_w = 6.8 Chengkung, 2006 M_w = 6.1 Taitung, and 2008 M_w = 5.4 Antung earthquakes can be clearly distinguished from other scattering radon data which appear to be related to smaller local earthquakes and aftershocks.

As shown in Fig. 11, radon decreased from background levels of 787 ± 42, 762 ± 57, and 700 ± 57 pCi/L to minima of 326 ± 9, 371 ± 9, and 480 ± 43 pCi/L prior to the 2003 M_w = 6.8 Chengkung, 2006 M_w = 6.1 Taitung, and 2008 M_w = 5.4 Antung earthquakes, respectively. Kuo et al. (2010b) recognized that the observed precursory minimum in radon concentration decreases as the local earthquake magnitude increases. Kuo et al. (2010b) also proposed an empirical correlation for local applications as follows.

$$C_w = 1063 - 110 M_w \qquad (2)$$

where C_w is the radon minimum in groundwater observed in well D1 during an anomalous decline, pCi/L; M_w is the earthquake magnitude. Eq. (2) did not take the initial stable radon concentration in groundwater precursory to each radon anomaly into account. Our observations in well D1 indicate that the initial stable radon concentration in groundwater precursory to each radon anomaly does vary occasionally. Eq. (2) will be improved by taking into account the initial stable radon concentration in groundwater precursory to each radon anomaly.

Based on radon phase behavior and rock dilatancy, Kuo et al. (2006, 2010a) developed a mechanistic model correlating the observed decline in radon with the volumetric strain change. The model consists of two parts, i.e., the radon-volatilization model and the rock-dilatancy model. The radon-volatilization model can be expressed as follows.

$$C_0 = C_w (H \times S_g + 1) \tag{1}$$

where C_0 is initial radon concentration in groundwater precursory to each radon anomaly, pCi/L; C_w is the radon minimum in groundwater observed in well D1 during an anomalous decline, pCi/L; S_g is gas saturation, fraction; H is Henry's coefficient for radon at formation temperature (60 ℃), dimensionless. The radon-volatilization model correlates the radon decline to the gas saturation for a given fracture porosity.

The rock-dilatancy model can be expressed as follows.

$$d\varepsilon \cong \phi \, S_g \tag{3}$$

where $d\varepsilon$ is volumetric strain, fraction; ϕ is initial fracture porosity before rock dilatancy, fraction; S_g is gas saturation, fraction. The rock-dilatancy model correlates the volumetric strain to the gas saturation for a given fracture porosity.

Combining the radon volatilization and rock dilatancy models, equations (1) and (3), the groundwater radon concentrations can be correlated to the strain changes associated with earthquake occurrences as follows.

$$d\varepsilon \cong \frac{\phi}{H} \left(\frac{C_0}{C_w} - 1 \right) \tag{4}$$

where $\left(\dfrac{C_0}{C_w} - 1 \right)$ is normalized radon decline, dimensionless. The Henry's coefficients (H) at formation temperature (60 ℃) is 7.91 for radon (Clever, 1979). Given an average fracture porosity of 0.00003 for the Antung hot spring, Eq. (4) can be used to calculate the crust strain.

Using the radon minima precursory to the 2003, 2006, and 2008 quakes, the calculated crust –strain and observed dimensionless radon-decline are plotted as a function of earthquake magnitude in Fig. 12. The best-fitting straight line is obtained by means of the least-square method with a high value of the sample correlation squared regression coefficient (i.e., R^2 = 0.9802). The regressed equations are as follows.

$$d\varepsilon = 2.5893 M_w - 12.0948 \tag{5}$$

$$\left(\frac{C_0}{C_w} - 1 \right) = 0.6827 M_w - 3.189 \tag{6}$$

where C_0 is initial radon concentration in groundwater precursory to each radon anomaly, pCi/L; C_w is the radon minimum in groundwater observed in well D1 during an anomalous decline, pCi/L; M_w is the earthquake magnitude; $d\varepsilon$ is volumetric strain, fraction. Eq. (6) would be quite useful locally in predicting earthquake magnitude nearby the Chihshang fault from the radon minimum observed in well D1 during an anomalous decline.

Three precursory radon minima associated with nearby large and moderate earthquakes have been recorded from the same monitoring well (D1). The same v-shaped pattern recognized in all the three recurrent radon anomalies and the threshold concentration are useful for the early warning of potentially disastrous earthquakes ($M_w > 6.0$) in the southern segment of coastal range and longitudinal valley of eastern Taiwan.

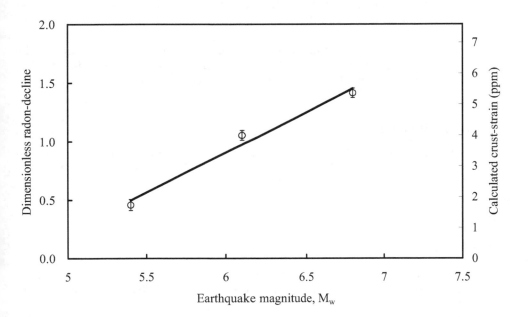

Fig. 12. Calculated crust-strain ($d\varepsilon$) and observed radon-decline ($\frac{C_0}{C_w} - 1$) at well D1 that occurred on December 10, 2003, April 1, 2006, February 17, 2008 as a function of earthquake magnitude (M_w). Radon-concentration errors are ±1 standard deviation.

6. Conclusions

Since July 2003, we have recorded three recurring radon anomalies (precursory to the 2003 M_w = 6.8 Chengkung, 2006 M_w = 6.1 Taitung, and 2008 M_w = 5.4 Antung earthquakes) at well D1, located at the Antung hot spring. The local geological conditions near the Antung hot spring with well D1 situated in a fractured aquifer surrounded by ductile mudstone and the in-situ volatilization of groundwater radon were attributed for causing the recurrent radon anomalies precursory to the nearby large and moderate earthquakes. The following conclusions can be drawn from this study.

1. Radon anomalous declines in groundwater consistently recorded prior to local large and moderate earthquakes near the Antung hot spring in eastern Taiwan provide the reproducible evidence to catch radon precursors under suitable geological conditions.
2. "A low-porosity fractured aquifer surrounded by ductile formation in a seismotectonic environment" is a suitable geological site to consistently catch precursory declines in groundwater radon and dissolved gases prior to local large and moderate earthquakes.
3. Radon partitioning into the gas phase (the mechanism of in-situ radon volatilization) may explain the radon anomalous declines in groundwater consistently recorded prior to local large and moderate earthquakes near the Antung hot spring in eastern Taiwan
4. The observed precursory minimum in radon concentration decreases as the earthquake magnitude increases. The observed relationship between radon minima and earthquake magnitude provides a possible means to forecast local disastrous earthquakes.

7. Acknowledgments

Supports by the National Science Council (NSC-96-2116-M-006-012, NSC-96-2738-M-006-004, NSC-97-2745-M-006-001, NSC-98-2116-M-006-016, and NSC-99-2116-M-006-019) and Central Geological Surveys (98-5226904000-05-02 and 99-5226904000-05-02), Radiation Monitoring Center, and Institute Earth Sciences of Academia Sinica of Taiwan are appreciated.

8. References

Angelier, J.; Chu, H. T.; Lee, J. C. & Hu, J. C. (2000). Active faulting and earthquake hazard: The case study of the Chihshang fault, Taiwan. *Journal of Geodynamics*, Vol. 29, pp. 151-185, ISSN 0264-3707

Brace, W. F.; Paulding, B. W. Jr. & Scholz, C. (1966). Dilatancy in the fracture of crystalline rocks. *Journal of Geophysical Research*, Vol. 71, pp. 3939-3953, ISSN 2156-2202

Chen, W. S. & Wang, Y. (1996). Geology of the Coastal Range, eastern Taiwan, In: *Geological Series of Taiwan 7*, F. C. Chien, Central Geological Survey, ISBN 957-00-6978-3, Taiwan

Clever, H. L. (1979). Krypton, Xenon and Radon-Gas Solubilities, In: *Solubility Data Series 2*, H. L. Clever & R. Battino, Pergamon Press, ISBN 0-08-022352-4, Oxford, UK

Hauksson, E. (1981). Radon content of groundwater as an earthquake precursor: Evaluation of worldwide data and physical basis. *Journal of Geophysical Research*, Vol. 86, pp. 9397-9410, ISSN 2156-2202

Hsu, T. L. (1962). Recent faulting in the Longitudinal Valley of eastern Taiwan. *Memoir of the Geological Society of China*, Vol. 1, pp. 95-102, ISSN 0578-1825

Igarashi, G.; Saeki, S.; Takahata, N.; Sumikawa, K.; Tasaka, S.; Sasaki, Y.; Takahashi, M. & Sano, Y. (1995). Ground-water radon anomaly before the Kobe earthquake in Japan. *Science*, Vol. 269, pp. 60-61, ISSN 1095-9203

Kuo, M. C. Tom; Fan, K.; Kuochen, H. & Chen, W. (2006). A mechanism for anomalous decline in radon precursory to an earthquake. *Ground Water*, Vol. 44, pp. 642-647, ISSN 1745-6584

Kuo, T.; Lin, C.; Chang, G.; Fan, K.; Cheng, W. & Lewis, C. (2010a). Estimation of aseimic crustal-strain using radon precursors of the 2003 M 6.8, 2006 M 6.1, and 2008 M 5.0 earthquakes in eastern Taiwan. *Natural Hazards*, Vol. 53, pp. 219-228, ISSN 1573-0840

Kuo, T.; Su, C.; Chang, C.; Lin, C.; Cheng, W.; Liang, H.; Lewis, C. & Chiang, C. (2010b). Application of recurrent radon precursors for forecasting large earthquakes (Mw > 6.0) near Antung, Taiwan. *Radiation Measurements*, Vol. 45, pp. 1049-1054, ISSN 1350-4487

Liu, K. K.; Yui, T. F.; Yeh, Y. H.; Tsai, Y. B. & Teng, T. L. (1985). Variations of radon content in groundwaters and possible correlation with seismic activities in northern Taiwan. *Pure and Applied Geophysics*, Vol. 122, pp. 231-244, ISSN 1420-9136

Lee, J. C.; Angelier, J.; Chu, H. T.; Hu, J. C.; Jeng, F. S. & Rau, R. J. (2003). Active fault creep variations at Chihshang, Taiwan, revealed by creep meter monitoring, 1998-2001. *Journal of Geophysical Research*, Vol. 108, pp. 2528-2548, ISSN 2156–2202

Lowry, D. (1991). Measuring low radon levels in drinking water supplies. *Journal American Water Works Association*, Vol. 83, pp. 149-153, ISSN 1551-8833

Noguchi, M. (1964). Radioactivity measurement of radon by means of liquid scintillation fluid. *Radioisotope*, Vol. 13, pp. 362-366, ISSN 0033-8303

Noguchi, M. & Wakita, H. (1977). A method for continuous measurement of radon in groundwater for earthquake prediction. *Journal of Geophysical Research*, Vol. 82, pp. 1353-1357, ISSN 2156–2202

Nur, A. (1972). Dilatancy, pore fluids, and premonitory variations of ts/tp traval times. *Bulletin of the Seismological Society of America*, Vol. 62, pp. 1217-1222, ISSN 1943-3573

Prichard, H. M. & Gesell, T. F. (1977). Rapid measurements of 222Rn concentrations in water with a commercial liquid scintillation counter. *Health Physics*, Vol. 33, pp. 577-581, ISSN 1538-5159

Prichard, H. M.; Venso, E. A. & Dodson, C. L. (1992). Liquid-Scintillation analysis of 222Rn in water by alpha-beta discrimination. *Radioactivity and Radiochemistry*, Vol. 3, pp. 28-36, ISSN 1045-845X

Scholz, C. H.; Sykes, L. R. & Aggarwal, Y. P. (1973). Earthquake prediction: A physical basis. *Science*, Vol. 181, pp. 803-810, ISSN 1095-9203

Teng, T.L. (1980). Some recent studies on groundwater radon content as an earthquake precursor. *Journal of Geophysical Research*, Vol. 85, pp. 3089-3099, ISSN 2156–2202

Toutain, J. P. & Baubron J. C. (1999). Gas geochemistry and seismotectonics: a review. *Tectonophysics*, Vol. 304, pp. 1–27, ISSN 0040-1951

Wakita, H.; Nakamura, Y.; Notsu, K.; Noguchi, M. & Asada, T. (1980). Radon anomaly: A possible precursor of the 1978 Izu-Oshima-kinkai earthquake. *Science*, Vol. 207, pp. 882-883, ISSN 1095-9203

Yu, S. B. & Kuo, L. C. (2001). Present-day crustal motion along the Longitudinal Valley Fault, eastern Taiwan. *Tectonophysics*, Vol. 333, pp. 199-217, ISSN 0040-1951

8

Radon as Earthquake Precursor

Giuseppina Immè and Daniela Morelli

Dipartimento di Fisica e Astronomia Università di Catania - INFN Sezione di Catania
Italy

1. Introduction

Earthquake predictions are based mainly on the observation of precursory phenomena. However, the physical mechanism of earthquakes and precursors is at present poorly understood, because the factors and conditions governing them are so complicated. Methods of prediction based merely on precursory phenomena are therefore purely empirical and involve many practical difficulties.

A seismic precursor is a phenomenon which takes place sufficiently prior to the occurrence of an earthquake. These precursors are of various kind, such as ground deformation, changes in sea-level, in tilt and strain and in earth tidal strain, foreshocks, anomalous seismicity, change in b-value, in microsismicity, in earthquake source mechanism, hypocentral migration, crustal movements, changes in seismic wave velocities, in the geomagnetic field, in telluric currents, in resistivity, in radon content, in groundwater level, in oil flow, and so on. These phenomena provide the basis for prediction of the three main parameters of an earthquake: place and time of occurrence and magnitude of the seismic event.

The most important problem with all these precursors is to distinguish signals from noise. A single precursor may not be helpful, the prediction program strategy must involve an integral approach including several precursors.

Moreover, in order to evaluate precursory phenomena properly and to be able to use them confidently for predictive purposes, one has to understand the physical processes that give rise to them. Physical models of precursory phenomena are classified in two broad categories: those based on fault constitutive relations, which predict fault slip behavior but no change in properties in material surrounding the fault, and those based on bulk rock constitutive relations, which predict physical property changes in a volume surrounding the fault. Nucleation and lithospheric loading models are the most prominent of the first type and the dilatancy model is of the second type.

During the past two decades efforts have been made to measure anomalous emanations of geo-gases in earthquake-prone regions of the world, in particular helium, radon, hydrogen, carbon dioxide. Among them radon has been the most preferred as earthquake precursor, because it is easily detectable.

Radon is found in nature in three different isotopes: ^{222}Rn, member of ^{238}U series, with an half life of 3.8 days, ^{220}Rn (also called thoron), member of ^{232}Th series, with an half life of 54.5 s and ^{219}Rn, member of ^{235}U series, with an half life of 3.92 s.

Owing to his longer half-life, the most important of them is ^{222}Rn, produced by ^{226}Ra decaying. After his production in soil or rocks, ^{222}Rn can leave the ground crust either by molecular diffusion or by convection and enters the atmosphere where his behavior and distribution are mainly governed by meteorological processes.

The radon decay products are radioactive isotopes of Po, Bi, Pb and Tl and they are easily attached to aerosol particles present in air. In table 1 are shown the principal decay characteristics of ^{222}Rn and ^{220}Rn, including properties of their respective parent radionuclides and their short-lived decay products.

Radionuclide	Half-life	Radiation	E_α(MeV)	E_γ(MeV)
^{226}Ra	1600 y	α	4.78(94.3%) 4.69 (5.7%)	0.186 (83,3%)
^{222}Rn	3.824 d	α	5.49(100%)	
^{218}Po	3.05 m	α	6.00(100%)	
^{214}Pb	26.8 m	β, γ		0.295 (19%) 0.352 (36%)
^{214}Bi	19.7 m	β, γ		0.609 (47%) 1.120 (15%) 1,760 (16%)
^{241}Po	164 μs	α	7.69 (100%)	
^{224}Ra	3.66 d	α	5.45 (6%) 5.68 (94%)	0.241(3.9%)
^{220}Rn	55 s	α	6.29 (100%)	
^{216}Po	0.15 s	α	6.78 (100%)	
^{212}Pb	10.64 h	β, γ		0.239 (47%) 0.300 (3.2%)
^{212}Bi	1.01 h	α, β, γ	6.05 (25%) 6.09 (10%)	0.727 (11.8%) 1.620 (2.8%)
^{212}Po	298 ns	α	8.78 (100%)	
^{208}Tl	3.05 m	β, γ		0.511 (23%) 0.583 (86%) 0.860 (12%) 2.614 (100%)

Table 1. Principal decay Characteristics of ^{222}Rn and ^{220}Rn

The release of radon from natural minerals has been known since 1920's (*Spitsyn, 1926*) but its monitoring has more recently been used as a possible tool for earthquake prediction, because the distribution of soil-gas radon concentration is closely related to the geological structure, fracture, nature of rocks and distribution of sources. Therefore, surveying of radon concentration can prospect fracture trace, earthquake forecast, environment monitoring, etc.

2. Radon production and transport

The production of ^{222}Rn depends on the activity concentrations of ^{226}Ra in the earth's crust, in soil, rock and water.

When radium decays in a mineral substance, the resulting radon atoms must first emanate from the grains into the air-filled pore space. The fraction of radon that enters the pores, commonly known as *emanation fraction*, consists of two components due to recoil and diffusion mechanisms. Since the diffusion coefficient of gases in solid materials is very low, it is assumed that the main portion of the emanation fraction comes from the recoil process. From the alpha decay of radium, radon atoms possess sufficient kinetic energy (86 keV) to move from the site where radon is generated. The range of ^{222}Rn is between 20 to 710 nm in common materials, 100 nm for water and 63 μm for air. (*Sabol et al., 1995*)

The emanation fraction can be strongly influenced by water content in the material, increasing with soil moisture, up to saturation in the normal range of soil moisture content. A representative estimate of the fraction of radon that leaves solid grains is 25%.

The increase in the emanation fraction can be explained by the lower recoil range of radon atoms in water than in air. A radon atom entering a pore that is fully or partially filled with water has a very good chance of being stopped by the water in the pore. Generally, the presence of water increases the emanation fraction, but this trend may show a saturation effect or the effect may even later reverse as the water content becomes greater.

In addition to the moisture effect, dependence of the emanation fraction on grain size and temperature has also been observed. Small grain size soils, such as clay, display maximum emanation at about 10%-15% water content. The ratio of the maximum emanation fraction to that of a dry sample also decreases as the grain size increases. A rise in temperature also causes an increase in the emanation fraction, which is probably due to the reduced adsorption of radon.

Different types of soil show different emanation fractions for ^{222}Rn, which are generally in the range 0.01-0.5 (*Sabol et al., 1995*).

Some emanated radon atoms, after their penetration trough the pore of a material, may finally reach the surface before decaying. Radon behaves as a gas and its movement in material follows some well-known physical laws. There are essentially two mechanisms of radon transport in material: (1) molecular diffusion and (2) forced advection.

In diffusive transport, radon flows in a direction opposite to that of the increasing concentration gradient. Fick's law describes this process. Expressions for the radon fluence rate, in Bq m^{-2} s^{-1}, can be derived for specified geometric conditions.

If one assumes the earth as a semi-infinite homogeneous material, with density ρ and porosity ε, the fluence rate J_D of radon emerging at the earth surface can be given by (*Sabol et al., 1995*):

$$J_D = C_{Ra}\lambda_{Rn}f\rho\left[\frac{D_e}{\lambda_{Rn}\varepsilon}\right]^{0.5} \tag{1}$$

where C_{Ra} is the activity concentration of ^{226}Ra in earth material (Bq/kg); λ_{Rn} is the decay constant of ^{222}Rn (2.1 $\times$ 10^{-6} s^{-1}); f and D_e are the emanation fraction and the effective diffusion coefficient for earth material (m^2/s) respectively.

After crossing soil-air interface radon exhales into the atmosphere. The exhalation rate, that is the amount of radon activity released from the surface, depends on meteorological parameters. In particular the exhalation of radon is positively correlated with moisture content, temperature and wind speed and negatively with pressure, so that these factors must be considered in the determination of exhalation rates in environmental measurements. Since the main mechanism governing the entry of radon into the atmosphere from the surface of the earth is diffusion, the radon fluence rate can be calculated by using appropriate parameters in equation (1). Representative values of these parameters and $C_{Ra} = 40$ Bq m^{-3} yield $J_D = 0.026$ Bq m^{-2} s^{-1} which is quite close to the average value experimentally obtained for some regions (*Sabol et al. ,1995*).

2.1 Theory of radon diffusion
In order to understand how radon anomalies could be correlated to geodynamic events radon transport mechanisms in soil must be considered. Different models to describe radon diffusion have been proposed. In this section we will give a brief review.

2.1.1 Plate sheet model
One of the most reliable models to describe radon diffusion is the plane sheet model. The molecular diffusion is considered in only one direction and, for any stable element, can be described by Fick's second law as follows (*Gauthier et al.,1999*):

$$\frac{\partial C}{\partial t} = D \frac{\partial^2 C}{\partial z^2} \tag{2}$$

where C is the concentration of the element and D the diffusion coefficient along z. This equation admits a solution $C(z, t)$ which is constrained by the initial and boundary conditions ($C = C_0$ at $t = 0$ and $-a < z < a$; $C = 0$ at $t > 0$ and $z = +a$):

$$C(z,t) = \frac{4C_0}{\pi} \sum_{n=0}^{\infty} \left\{ \left(\frac{(-1)^n}{2n+1} \right) \times \cos\left(\frac{(2n+1)\pi z}{2a} \right) \times \exp\left(\frac{-D(2n+1)^2 \pi^2 t}{4a^2} \right) \right\} \tag{3}$$

where a is the half-width of the slab.
In order to take into account radioactivity, equation (3) has to be modified for radon by adding a production term from its parent ^{226}Ra and a decay term, which leads to:

$$\frac{\partial C_{Rn}}{\partial t} = \lambda_{Ra} C_{Ra} - \lambda_{Rn} C_{Rn} + D \frac{\partial^2 C_{Rn}}{\partial z^2} \tag{4}$$

where C_{Ra} and C_{Rn} represent the concentrations (in atoms $\cdot$ g^{-1}) and λ_{Ra} and λ_{Rn} the decay constants of ^{226}Ra and ^{222}Rn, respectively.
Defining the function $K(z, t)$ as:

$$K(z,t) = \left(C_{Rn}(z,t) - \left(\frac{\lambda_{Ra}}{\lambda_{Rn}} C_{Ra} \right) \right) \exp(\lambda_{Rn} t) \tag{5}$$

and introducing $K(z, t)$ in equation (4) gives:

$$\frac{\partial K}{\partial t} = D \frac{\partial^2 K}{\partial z^2} \tag{6}$$

which is the Fick's second law expressed for the function K(z, t). Nevertheless the solution of equation (6) cannot be merely obtained by combining the solution of the general Fick's second law (2) with the substitution (5) because the two functions $K(z, t)$ and $C_{Rn}(z, t)$ do not admit the same initial and boundary conditions. These conditions are for $C_{Rn}(z, t)$:

$$C_{Rn}(z,0) = C_{RnEq} = \frac{\lambda_{Ra}}{\lambda_{Rb}} C_{Ra} \qquad \text{for} -a < z < a, t = 0$$

$$C_{Rn}(z,t) = 0 \qquad \text{for} \quad z = -a, z = a \tag{7}$$

(the atmosphere is considered as a reservoir of concentration $C = 0$) which means for $K(z, t)$:

$$K(z,0) = 0 \qquad \text{for} -a < z < a, t = 0$$

$$K(z,t) = -C_{RnEq} \exp(\lambda_{Rn}t) \qquad \text{for} \quad z = -a, z = a \tag{8}$$

Fick's law is usually solved for plane sheet geometry by separation of variables but this method is unsuccessful for such initial and boundary conditions. Several studies have been done for heat conduction in a slab having an initial zero temperature and surfaces maintained at the temperature $f(t) = V \exp(vt)$ (*Gauthier et al.,1999*), obtaining:

$$K(z,t) = -C_{RnEq} \exp(\lambda_{Rn}t) \frac{\cosh\left(z\sqrt{\frac{\lambda_{Rn}}{D}}\right)}{\cosh\left(a\sqrt{\frac{\lambda_{Rn}}{D}}\right)} +$$

$$+ \frac{4C_{RnEq}}{\pi} \sum_{n=0}^{\infty} \frac{(-1)^n \exp\left(\frac{-(2n+1)^2 \pi^2 Dt}{4a^2}\right)}{(2n+1)\left[1 + \left(\frac{4\lambda_{Rn}a^2}{(2n+1)^2 \pi^2 D}\right)\right]} \cos\frac{(2n+1)\pi z}{2a} \tag{9}$$

and therefore, combining with (6):

$$C_{Rn}(z,t) = C_{RnEq} - C_{RnEq} \frac{\cosh\left(z\sqrt{\frac{\lambda_{Rn}}{D}}\right)}{\cosh\left(a\sqrt{\frac{\lambda_{Rn}}{D}}\right)} +$$

$$+ \frac{4C_{RnEq}}{\pi} \sum_{n=0}^{\infty} \frac{(-1)^n \exp\left(-\left(\frac{(2n+1)^2 \pi^2 D}{4a^2} + \lambda_{Rn}\right)t\right)}{(2n+1)\left[1 + \left(\frac{4\lambda_{Rn}a^2}{(2n+1)^2 \pi^2 D}\right)\right]} \cos\frac{(2n+1)\pi z}{2a} \tag{10}$$

By multiplying by λ_{Rn} both sides of the equation (10), one obtains the activity of $C_{Rn}(z, t)$ given in equation:

$$C_{Rn}(z,t) = C_{Ra} - C_{ra} \frac{\cosh\left(z\sqrt{\frac{\lambda_{Rn}}{D}}\right)}{\cosh\left(a\sqrt{\frac{\lambda_{Rn}}{D}}\right)} +$$

$$+ \frac{4C_{Ra}}{\pi} \sum_{n=0}^{\infty} \frac{(-1)^n \exp\left(-\left(\frac{(2n+1)^2 \pi^2 D}{4a^2} + \lambda_{Rn}\right)t\right)}{(2n+1)\left[1+\left(\frac{4\lambda_{Rn}a^2}{(2n+1)^2 \pi^2 D}\right)\right]} \cos\frac{(2n+1)\pi z}{2a} \tag{11}$$

where C_{Rn} and C_{Ra} represent the activity of ^{222}Rn and ^{226}Ra, respectively.

2.1.1 Infinite source model

In another earth model an infinite source C_0 is overlain by an overburden of thickness h, where no radon source exists. In this case the radon transportation equation in the overburden, where radon production rate is zero, can be written as (Wattanikorn et al,1998):

$$\frac{d^2C}{dz^2} + \frac{v}{D}\frac{dC}{dz} - \frac{\lambda_{Rn}}{D}C = 0 \tag{12}$$

where C is the radon concentration at depth z, v is the gas flow velocity; D is the diffusion coefficient of radon, and λ_{Rn} is the decay constant. The solution of (12) is:

$$C_{Rn} = C_{Rn0} \exp\left[\frac{v(h-z)}{2D}\right] \frac{\sinh\left[\sqrt{\left(\frac{v}{2D}\right)^2 + \frac{\lambda_{Rn}}{D}}\,z\right]}{\sinh\left[\sqrt{\left(\frac{v}{2D}\right)^2 + \frac{\lambda_{Rn}}{D}}\,h\right]} \tag{13}$$

3. Radon measurements

In general Radon measurements can be performed in continuous, integrating or discrete mode, regarding the time duration of measurement, and by using passive devices, when Radon enters the detection system by natural diffusion, or active technique, when gas is pumped in the device, that require electric power.

Some types of the most used detectors for in-soil radon measurements are the following:

a. **Solid State nuclear track detectors**: the most used SSNTD are Cr-39 type or LR-115 one. They are particularly sensitive to alpha particles that, passing trough, produce tracks visible in optical microscope after chemical etching. The main advantages of this kind of detectors, especially for the first type, is that they are cheap, are sensitive only to alpha particles, are unaffected by humidity, low temperature, moderate heating and light. Moreover these passive devices don't need electrical power supply.

b. **Electret detector**: an electret is a dielectric material that exhibits a permanent electrical charge. The particles from Radon decay produce ions within the device that determine changing in the total charge of the electret. This kind of detector offers several advantages: possibility to store information over relatively long period, independence from moisture in its envelop and ease to read. The main problems are linked to its response curve that does not cover efficiently the very low and very high doses and its sensitivity also to gamma radiations.

c. **Activated charcoal**: this type of detectors is based on the capability of the charcoal to adsorb Radon gas. The analysis is carried out by means of the gamma spectrometry of the Radon products. However with this kind of device measurements can be performed only for 3 – 5 days and they are affected by humidity.

d. **Thermoluminescent detector**: Radon is allowed to enter the detection device volume containing the TLD. A metallic plate, placed at short distance in front of the TLD, can be electrically charged for a better collection efficiency. Radon daughters deposited on the plate decay producing energy storage in the TLD. After appropriate exposition, the TLD is recovered and read out in a TLD apparatus.

e. **Scintillation detector**: the most widely used is the ZnS(Ag) scintillation cell for grab sampling. It is a metal container internally coated with silver activated zinc solphide. Light photons are detected, resulting from the interaction of the alpha particles from radon decaying. For counting the photons, the scintillation cell is coupled to a photomultiplier.

In the last years active devices have been used for continuous measurements of in soil Radon gas. They use prevalently detectors as ionization chamber or silicon detectors. The devices have a probe placed in the soil at a certain depth, the gas Radon enters into the detection chamber or by means of a pump with a fixed flow rate or they can be placed inside the soil and the gas enters into the detection chamber via natural diffusion. This kind of measurements need power supply, not always available in active fault areas, but in the last years the detection systems have been implemented with solar panels, overcoming the problem. These devices have more performance respect to the previous ones because they allow continuous measurements and on-line reading by means of remote data transfer and so they allow to monitor continuously the Radon temporal trend.

Accordingly, the choice among the different possibilities can be guided by the particular interest in radon measurements, whether in time-dependent or in space-dependent variations of the concentrations. In particular, spot measurements (with portable detectors) of soil–gas Radon are useful for the quick recognition of high emission sites to be later monitored for Radon variations in time. SSNTD allow for the temporal monitoring of a relatively large number of sites, but cannot distinguish short-term changes due to their long integration times. Continuous monitoring probes are optimal for defining detailed changes in soil–gas Radon activities, but are expensive and can thus be used to complete the information acquired with SSNTD in a network of monitored sites.

4. Origin and mechanisms of radon anomalies

Most of the researchers define radon anomaly as the positive deviation that exceeds the mean radon level by more than twice the standard deviation.

The origin and the mechanisms of the radon anomalies and their relationship to earthquakes are yet poorly understood, although several in-situ and laboratory experiments have been performed and mathematical modelings have been proposed. The radon observed in case of anomalies correlated with geophysical events may be considered as having two possible origins. Either it is produced in depth origin or, once produced locally, it is displaced by other interstitial fluids whose motion is triggered by geodynamical events. Both possibilities have been discussed so far but the local origin hypothesis seems to be the most reliable, sustained by experiments too. In order to relate radon anomalies to earthquake occurrence, several scenarios have been proposed. Accordingly to the dilatancy-diffusion model (*Scholz, 1973; Planinic et al, 2001*) the radon anomalies could be related to mechanical crack growth in the volume of dilatancy or to changes in groundwater flow. Consequently either opening of new cracks, widening or closing of old cracks or redistribution of open and closed cracks can happen. In dry rocks opening or closing of cracks will lead to significant changes of the diffusion coefficient of radon. Volumetric changes in the rock will also lead to a subsurface gas flow and therefore to an additional radon transport. If the new open cracks are filled with water the increased water-rock interface leads to an increase in the transfer of radon from the rock matrix to the water. If water filled cracks close, the water will be compressed to another subsurface volume where the emanation from the rock to the water may be different. All these effects result in pressure and water level variations of the relevant aquifer. This also can lead to changes in the mixing ratios for the water which can be observed at the earth's surface. Finally gas flows can also move some groundwater and again all previously discussed mechanisms which are consequences of the redistribution of water in the earth's crust can take effects. This scenario has the drawback that an unreasonably large change in stress or strain is required far away from the epicentre.

An alternative mechanism is the stress corrosion theory, first proposed by *Anderson and Grew (1977)*. It attributes the radon anomalies to slow crack growth controlled by stress corrosion which should precede any mechanical cracking in wet environment. According to the mechanism of stress corrosion radon anomalies may depend on strain rate and local conditions such as rock type, porosity, elasticity, pattern of micro-cracks, degree of saturation, temperature, stress intensity factor and hygroscopic properties.

If the local parameters of the rocks are assumed to be responsible for the radon anomalies an explanation is needed to justify how very small changes in the stress field can produce such effects. From the theory of the earthquake preparation process it could be derived that in a region where the stress reaches a level which is not very far away from rock failure, very small changes in the stress fields result in considerable changes in certain rock parameters. If this theory holds, earthquake sensitivity could be expected only in areas which are highly pressed, for example near fault zone systems, not necessarily seismic active.

According to another kind of mechanism, the compression mechanism proposed by *King (1978)*, the anomalous radon concentration may be due to an increase in crustal compression, impending an earthquake, that squeezes out the soil-gas into the atmosphere at an increased rate. Radon anomalies have been observed at large distance from the earthquake epicentre, resulting from changes in the immediate vicinity of recording station, rather than at the distant focal region. This is accomplished if it is assumed that changes in stress or strain are propagated from the rupture zone to the radon station, leading to variations also in porosity, emanating power or flow rate of the local groundwater, near the radon monitoring station.

When the diffusion constant of radon in a soil of average porosity and moisture content is considered, the calculation shows that radon cannot be detected at a distance larger than a few meters. When radon is pumped by an upward moving carrier, whose motion is of the order of a few microns per second, it of course increases its concentration near the monitoring station while the underground is depleted in radon. Since the half-life of radon is of 3.82 d, it would take a time larger than what is observed actually for those lower parts to supply by radioactive decay only. A much longer radon gas column than the few meters involved by mere diffusion should be involved. However, the most of the radon would decay away before reaching the detection system. The motion of pore fluids would first of all increase the radon concentration temporarily and then exhaust the available resources for further increase.

4.1 Radon anomaly shapes

When radon concentrations are measured in continuous mode for a long time and with a time resolution of at least one hour, it is possible to classify the observed radon anomalies according to different trends. Typically two shapes can be considered that *Friedman, 1991* classified group A and group B anomalies. Probably the physical process is different for the two groups of anomalies.

The group A shows a very slow increase (or decrease) of the radon concentration with a rate less than 0.1% per hour. This kind of anomaly can be linked to a continuous increase in stress, until the rock fracture occurency.

The second group B is characterized by a fast increase (or decrease) in the radon concentration with a rate of about 1% per hour. Often a fast increase is followed by a rather constant radon concentration. Sometimes anomaly spikes with fast radon change immediately followed by a fast change in the opposite direction. These two kind of B anomalies could be linked to different physical processes or simply to different time scales (*Friedmann, 1991*).

The B-type anomalies can be a local effect, which depends on certain local parameters, or they can be an epicentral effect. In this case the epicentral area must be supposed as origin of the fast change in stress.

For spike anomalies the maximum velocity of the radon concentration change is:

$$V = \frac{1}{\delta C}\left(\frac{dC_{Rn}}{dt}\right)_{max} \geq \frac{1}{\tau} \tag{14}$$

where C_{Rn} is radon concentration
δC_{Rn} the difference between radon concentration before and after the fast change
τ the time of the fast change (in hours)

From data of V and the epicentral distance d a very rough correlation can be found for $d>70$km, according to a relation of the form:

$$log(V) = -2log(d) + 4 \quad [V] = hours^{-1}, [d] = km \tag{15}$$

Thus by considering V from an observed anomaly a rough estimation of the distance from the epicentre can be made.

5. Forecasting relations

Earthquake prediction means to forecast place, time and magnitude of an earthquake. From the analysis of a wide variety of radon data available from different countries and earthquakes with $M<3$, *Rikitake* proposed an empirical relation between the time interval t between radon anomaly and earthquake occurrence and magnitude of an earthquake (*Rikitake, 1976*):

$$Log\ t = 0.76M - 1.83 \tag{16}$$

The relation was modified by *Fleischer* depending on the time interval: (*Rikitake, 1976*):

$$Log\ t = M - 2.16 \quad \text{for } 0{,}1 < t < 7 \text{ days} \tag{17}$$

$$Log\ t = 0.62M - 1.0 \quad \text{for } t > 7 \text{ days} \tag{18}$$

Starting from the radon diffusion equation and analyzing radon data from many countries , *Ramola et al (1988)* proposed an empirical relation to predict the magnitude of strong earthquakes ($M>5$) :

$$M = 2log(\lambda_{Rn}\Delta C_{Rn}/KT) - 15.26 \tag{19}$$

where ΔC_{Rn} is the anomalous variation of radon concentration, T rise time for radon anomaly and K is a constant (3.96×10^{-17}).

Several models were suggested in the past to evaluate the size of the area subject to changes in the tensional state. The models are based on assumption of homogeneity and isotropy of the ground or little heterogeneity around the focal zone.

In particular, *Dobrovolsky et al.(1979)* proposed some relations, taking into account an ellipsoidal inclusion with a 30% of heterogeneity with respect to the surrounding ground. He obtained the following relations that connect the magnitude M and the maximum distance R that the deformation can reach with the amplitude of the deformation E:

$$
\begin{aligned}
E &= \frac{10^{1.5M-9.18}}{R^3} & M &< 5.0 \\[2mm]
E &= \frac{10^{1.3M-8.19}}{R^3} & M &\geq 5.0
\end{aligned}
\tag{20}
$$

On the basis of these relationships deformations that can generate an anomaly were evaluated to be of the order of 10^{-8} (*Hauksson, 1981*).

Since radon anomalies seem to have a local origin, it is important to consider a relationship between the magnitude and the distance to the epicentre.

If the maximum possible distance d between the epicentre of a forthcoming earthquake and the spring which can be influenced by this earthquake is proportional to the volume of the pre-stressed lithosphere or to the energy of the earthquake respectively, a relation holds of the form (*Friedmann, 1991*):

$$M = a \bullet log(d) + b \quad a,b = const \tag{21}$$

From known relations between magnitude M and the volume of the focal zone the a-value can be determined to be about 2. *Dobrovolsky et al.(1979)* observed that precursory phenomena are not observed beyond the distance d, thus to estimate roughly the radius of the effective precursory manifestation zone, they proposed the formula:

$$d = 10\ exp\ 0.43\ M \tag{22}$$

where d is in km and M is the magnitude of the earthquake. It means that a magnitude 5 earthquake will be detected by means of precursory phenomena at a distance not greater than 142 km.

By collecting and analyzing radon anomaly data *Hauksson et and Goddard (1981)* found a similar relation. It is important that all these relations do not differ by more than 30% in d for $M \geq 4$. But the most interesting result is that all observed precursors are limited by a straight line which coincides practically with a computed deformation of 10^{-8}. Summarizing the results in only one formula it is possible to estimate the magnitude-limit Mmin for the possibility of detecting a precursor anomaly at a distance d (in Km) to be

$$M \geq Mmin = (2.3 \pm 0.2) log(d) - (0.4 \pm 0.3) \tag{23}$$

The constants in (23) may differ for different areas, however it is a good over all approximation. Of course we can expect that certain directions from the future epicentre are favored compared to others. The limit (23) must be seen as the limit for the favored directions.

Another relation was proposed by *Martinelli, 1992* for which:

$$M = 2.4 log d - 0.43 - 0.4 \tag{24}$$

While precursor time t (in days) related to the magnitude M and the epicentral distance d (in km) can be estimated as follows:

$$log dt = 0.63M + (-) 0.15 \tag{25}$$

Long term series analyses have revealed a relation between the amplitude and duration of the gaseous anomaly and the magnitude M of the expected earthquake (*Barsukov et al., 1984*):

$$M = K\sqrt{S} \tag{26}$$

where K is a correction factor and S is the area of the peak anomaly, thus the shape of the peak is a diagnostic parameter for the forthcoming seismic event.

6. Radon anomalies and earthquakes: Some cases

Several radon investigations have been carried out all over the world. Measurements of this gas both in soil and in groundwater have shown that spatial and temporal variations can provide information about geodynamical events.

In the following we report some examples of studies among the numerous ones performed around the world with the purpose to relate abnormal radon emission to seismic events.

The pioneering work on radon investigation in groundsoil was performed at an active fault zone for two years (*Hatuda, 1953*). Radon concentration in soil gas was measured and anomalous radon concentrations were reported before the strong earthquake (M=8) of Tonankai (December 1944, Japan).

Some years later Tanner (*Tanner, 1959*) evidenced the importance of the influence of the meteorological parameters on radon measurements and in 1964 he suggested that radon could be used as tracer to discover uranium deposits or to predict earthquakes (*Tanner, 1964*).

The first evidence of radon in groundwater as precursor of earthquakes was observed in Tashkent (*Ulomov, 1967*). The author observed that the radon concentration in a spring near Tashkent increased constantly before the M=5.2 earthquake on April 15, 1966.
Afterward many studies have been performed about radon anomalies and earthquakes.
In the following some examples are reported on ground radon monitoring in the most seismic regions in the world.

6.1 Japan

As already cited, studies performed by *Hatuda (Hatuda 1953)*, at an active fault zone evidenced anomalous radon concentration before the strong earthquake (M=8) of Tonankai.
Radon anomalies were recorded before the Nagano Prefecture earthquake (M= 6.8) on September 14, 1984 (*Hirotaka et al., 1988*). The authors observed a gradual increase in radon counts three months before the quake and a remarkable increase two weeks before the shock.
For about twenty years an extensive network of groundwater radon monitoring has been operated mainly by the University of Tokyo and the Geological Survey of Japan for the purpose of earthquake prediction in eastern Japan. In figure 1. a significant example of radon anomaly is reported (*Igarashi et al., 1995*). The authors performed radon concentration analysis in a well 17 m deep from November 1993 to March 1995 and observed stable radon concentration of 20 Bq/l at the end of 1993. The radon concentration started to increase gradually from October 1994 reaching 60 Bq/l on November 1994, three times that in the same period one year before. Furthermore, a sudden increase of radon concentration, recorded on 7 January was followed by a sudden decrease on 10 January, 7 days before an earthquake of magnitude 7.2. After the earthquake, the radon concentration returned to the pre-October 1994 levels. The main result of this example is that it is possible to observe strange behavior before an anomaly. This, for instance, as in this case, must be preceded by a continuous increasing in the background level till its manifestation. Naturally it depends on the geodynamical evolution of the area

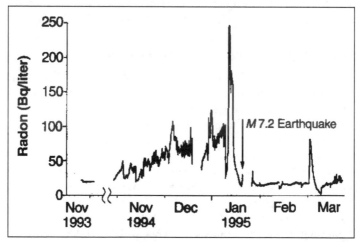

Fig. 1. Radon concentration data at the well in the southern part of Nishinomiya city, Japan [From . *Igarashi et al., 1995*]

6.2 India

In Bhatsadam, Maharashtra, India, major earthquakes occurred during August 1983 - July 1984. In that region radon concentration was measured by *Rastogi et al.(1986)*. They found an increase in radon concentration during March–April 1984 when seismicity was high enough. Precursory phenomena of radon in earthquake sequence were observed by *Rastogi et al. (1987)* and by other groups at the Osmansagar reservoir, Hederabad, India during January–February, 1982 (*Rastogi et al.,1987*). An earthquake with a magnitude of 3.5 occurred on January 14, 1982 with subsequent seismic events. There was an increase of radon concentration in soil gas during February due to those high seismic activities.

Singh et al. (1991) performed a daily radon monitoring in soil-gas in Amritsar from 1984 to 1987. They recorded radon anomalies before different earthquakes: June 1988 (M=6.8); April, 26, 1986 (M=5.7); July 1986 (M=3.8); Kangra earthquake March 1987 (M=7) and May 1987 (M= 5).

Virk and Singh (1994) carried out daily measurements of radon in soil-gas and groundwater at Palampur since 1989 and radon anomaly was recorded simultaneously in both soil- gas and groundwater. Weekly integrated data also showed abnormal radon behaviour during first week of October, 1991 at different recording stations. These recorded anomalies were correlated with an earthquake of magnitude 6.5 occurred in Uttarakashi area in October 1991.

6.3 Syria

Al-Hilal et al. (1998) recorded groundwater radon data for two years, during 1993 and 1994 at monthly intervals, from two selected monitoring sites of the northern extension of the Dead Sea Fault System. The results showed that measured radon concentrations fluctuate around the mean value, showing some variations with peak values, about two or three times the mean value, preceding some seismic events. It is possible to consider those anomalies related to changes in crustal strain and thereby to indicate a probable relation with the local seismicity. Nevertheless, the authors conclude that this does not necessarily means that it is possible to relate univocally these radon peaks to seismic event occurrence, but rather, it may indicate the possibility of using groundwater radon variations as a useful tool.

6.4 Turkey

In soil radon gas was monitored by *Friedmann et al. (1988)* in a network of five monitoring sites along 200 km at the North Anatolian Fault Zone, Bolu. They observed an increase in radon concentration during the strong earthquake (M=5.7) on July 5, 1983. In order to search some relation between earthquakes and radon concentration variations, more recently *Inceoz et al (2006)* performed a radon investigation at the North and East Anatolian fault system. They found that radon anomaly was quite significant in particular over the fault line but not away from this line.

Also the Aksehir fault zone was investigated, by *Baykara and Dogru (2006)* and *Yalim et al. (2007)*, trough radon measurements in well water. Although the observed radon levels could be related to several seismic activity that at the fault region occurred with high magnitude, the authors did not infer correlation between seismic activity and radon concentration.

Radon concentration in thermal water was investigated by *Erees et al. (2006,2007)* at two thermal springs at the Denizli basin site and significant radon anomalies were observed before earthquakes with magnitude between 3.8 and 4.8.

6.5 Italy

In the last ten years systematic studies on Radon as precursor of geophysical events have been carried out on Mt. Etna since 2001 (*Immè et al, 2005; Immè et al. 2006a, Immè et al, 2006b; La Delfa et al. 2007; La Delfa et al., 2088; Morelli et al. 2006, Morelli et al., 2011*). In particular two sites were investigated among the cropping up structural discontinuities, which lie along the NE-SW direction through the volcano. One site (*Biancavilla*) is in the SW flank, while the other one (*Vena*) is in the NE flank (circles in fig.2). Continuous monitoring was performed by using active systems with time resolution of 10 min. Capillary probes inserted into the soil at one meter depth, allowed to reduce influence from the meteorological parameters that were measured too.

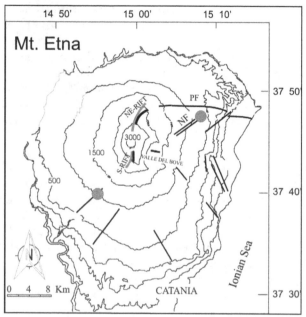

Fig. 2. Mt. Etna map– Circles indicate the sites where devices for continuous in soil gas Radon monitoring were positioned

Several studies conducted in tectonic areas evidenced relation to earthquakes of magnitude bigger than 3 (*Igarashi et al., 1995; Virk et al., 1994, Al-Hilal et al., 1998*). The etnean area is characterized by a big number of earthquakes, up to about thousands per day before an eruptive period (*Benina et al,1984; Patanè et al, 1995*), but with low magnitude (< 3) and rarely they exceed magnitude 4. Moreover Mt. Etna has a very complex structure, due to the occurrence of both tectonic and volcanic phenomena. Major results have been obtained respect to a possible link between radon concentration and volcanic activity. Nevertheless, some relations were also observed with seismic events as reported by *Immè et. al, 2005*, the data are referred to the period 2001-2002. Radon concentration values started to increase the 27th of October 2002, reached the maximum the 1st of November 2002 and the minimum the 3rd of November 2002. During this period several earthquakes of magnitude higher than 3 occurred, some of them reached values up to M= 4.5 (29/10/02 time 09:02:00 epicentral area of *Santa Venerina*).

It was observed that, as well as the radon raises the earthquake daily rate and strain release raise, correspondently at the eruption beginning.
A radon anomaly was recorded before the November 3rd event (M= 3.5), with epicentral zone close (less than 1 km) to the *Vena* Station (NE station), also associated to evident soil fractures.

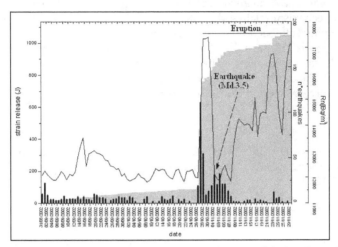

Fig. 3. Radon concentration (black line), daily earthquakes rate (black column bar) and strain release (grey histogram) measured in the period between 1st September 2002 and 30th November 2002 (Vena station).[*Immè et al., 2005*]

More recently a systematic radon investigation was extended to fault systems, in particular the Pernicana fault, one of the more active etnean fault, was chosen as first monitoring area. In particular, two different horizontal profiles, orthogonally to the main fault plane, were investigated. The first one was located at 1400m asl, the second one at 1370m asl (*Giammanco et al, 2009*). Each profile consisted of ten measurement points where CO_2 efflux values were also measured. Concentrations of ^{222}Rn were obtained by means of three different methodologies: passive, spot and continuous. The pattern of soil ^{222}Rn values measured in the two profiles is clearly similar: higher values were generally recorded on the up thrown side of the fault and the lowest values occurred generally close to the main fault plane. Differently to radon, higher CO_2 emissions were recorded on the fault plane. This behavior can be justified by the in-soil gas transport mechanism. In particular, along the main fault plane, advective transport of deep gases (CO_2, Rn) occurs because of the high ground fracturation and permeability. Near the surface, dilution of radon by CO_2 prevails, thus producing lower radon values.
This kind of investigations is useful to study the dynamics of the faults and the possible earthquake mechanisms.

7. Conclusion

From many years a lot of efforts have been done in order to improve in-situ radon data monitoring and analysis, technical methodologies and mathematical modeling, with the aim to reinforce the link between ground radon concentration anomalies and geodynamical events. Measurements of radon gas in soil and in ground water have been carried out all over the world and the results seem to indicate the radon as a good indicator of crustal activity such as earthquakes. However, the current literature describing the possible correlation between radon levels and earthquake activity uses such qualifying and caution words as possible, apparent, limited, could, sometimes, may be, and so on.

It is clear that in some cases there are precursor changes in radon levels, but that the causal relationship or mechanism relating these to earthquake activity is not yet well understood. Thus, even if some results seem to suggest that geodynamical events could influence radon concentrations, however, because of the complexity of its transport mechanism, the correlation needs more investigations in order to clearly and firmly established it.

Further contributions can be obtained from more extended continuous data recording, in particular near active faults, and from the comparison with other earthquake precursors.

8. References

Al-Hilal, M., Sbeinati, M.R. and Darawcheh, R. (1998) Radon variation and microearthquakes in western Syria. *Applied Radiation and Isotopes* Vol.49, Nos.1-2, pp. 117-123

Anderson, O.L., and Grew, P.C. (1977) Stress corrosion theory of crack propagation with application to geophysics. *Rev. Geophys. Space Phys.*, Vol. 15, 77-104.

Antsilevich, M.G. (1971). An attempt to forecast the moment of origin of recent tremors of the Tashkent earthquake through observations of the variation of radon. *Izvestiâ Akademii nauk Uzbekskoj* SSR 188-200.

Barsukov, V.l., Varshal, G.M., Garanin, A.B., and Serebrennikov, V.S. (1984). Hydrochemical Precursors of Earthquakes. *Earthquake Prediction, UNESCO, Paris,*169-180.

Baykara, O., Dogru, M. (2006). Measurements of radon and uranium concentration in water and soil samples from East Antolian active fault systems (Turkey). *Radiation Measurements* 41 (3), 362–367.

Benina, A., Imposa, S., Gresta, S., Patanè, G. (1984). Studio macrosismico e strutturale di due terremoti tettonici avvenuti sul versante meridionale dell'Etna, *Atti III convegno annuale del GNGTS*; 931-946

Dobrovolsky, I.P., Zubkov, S.I., Achkin, V.I. (1979). Estimation of the size of earthquakes preparation zone. *Paleoph.,*Vol.117, 1025-1044.

Erees, F.S., Yener, G., Salk, M., Ozbal, O. (2006). Measurements of radon content in soil gas and in the thermal waters in Western Turkey. *Radiation Measurements*, 41, 354–361.

Erees, F.S., Aytas, S., Sac, M.M., Yener, G., Salk, M. (2007). Radon concentrations in thermal waters related to seismic events along faults in the Denizli Basin, Western Turkey. *Radiation Measurements*, 42, 80–86.

Friedmann, H., Aric, K., Gutdeutsch, R., King, C.Y., Altay, C., Sav, H. (1988). Radon measurements for earthquake prediction along the North Anatolian Fault Zone: a progress report. *Tectonophysics* 152 (3–4), 209–214.

Friedmann, H. (1991) Selected problems in Radon measurement for earthquake prediction *Proceedings of the Second workshop on Radon Monitoring in Radioprotection, Environmental and/or Earth Science*, Furlan, G. and Tommasino, L. (Ed.) World Scientific.307-316.

Gauthier, P-J. and Condomines, M. (1999). ^{210}Pb - ^{226}Ra radioactive disequilibria in recent lavas and radon degassing: inferences on the magma chamber dynamics at Stromboli and Merapi volcanoes. *Earth and Planetary Science Letter,* Vol. 172, 111-126.

Giammanco, S., Immè, G., Mangano, G., Morelli, D., Neri, M. (2009). Comparison between different methodologies for detecting Radon in soil along an active fault: the case of the Pernicana fault system, Mt. Etna(Italy). *Applied radiation and Isotopes* 67, 178 -185.

Hauksson, E. (1981). Radon content of groundwater as an earthquake precursor: evaluation of worldwide data and physical basis. *Journal of geophysical research,* Vol. 86, 9397-9410.

Hauksson, E., Goddard J.G (1981). Radon earthquake precursor studies in Iceland. *J. Geophys. Res.* , Vol.86, No. B8, 7037-7054

Hatuda, Z. (1953). Radon content and its change in soil air near the ground surface. *Memoirs of the College of Science,* University of Kyoto, Series B 20, 285–306.

Hirotaka, U., Moriuchi, H., Takemura, Y., Tsuchida, H., Fujii, I., Nakamura, M. (1988). Anomalously high radon discharge from the Atotsugawa fault prior to the western Nagano Prefecture earthquake (m 6.8) of September 14, 1984. *Tectonophysics* 152 No 1–2, 147–152.

Igarashi, G., Saeki, S., Takahata, N., Sumikawa, K., Tasaka, S., Sasaki, Y., Takahashi, M., Sano ,Y. (1995). Ground-water radon anomaly before the Kobe earthquake in Japan *Science* Vol. 269, 60-61.

Imamura, G. (1947). Report on the observed variation of the Tochiomata hot spring immediately before the Nagano earthquake of july 15, 1947, *Kagaku,* 11, 16-17

Immè, G., La Delfa, S., Lo Nigro, S., Morelli D., Patanè, G. (2005). Gas Radon emission related to geodynamic activity of Mt. Etna. *Annals of Geophysics,* 48 N.1, 65-7.

Immè, G., La Delfa, S., Lo Nigro, S. , Morelli D., Patanè, G. (2006a) Soil Radon concentration and volcanic activity of Mt. Etna before and after the 2002 eruption. *Radiation Measurements* Vol.41, 241-245.

Immè, G., La Delfa, S., Lo Nigro, S., Morelli D., and Patanè, G. (2006b). Soil Radon monitoring in NE flank of Mt. Etna (Sicily), *Applied Radiation and Isotopes,* Vol.64, 624-629.

Inceoz, M., Baykara, O., Aksoy, E., Dogru, M. (2006). Measurements of soil gas radon in active fault systems: a case study along the North and East Anatolian fault systems in Turkey. *Radiation Measurements* 41 (3), 349–353.

King, C.Y. (1978). Radon emanation on San Andreas Fault. *Nature,* Vol. 271, 576-519.

La Delfa, S., Immè, G., Lo Nigro, S., Morelli, D., Patanè, G., Vizzini, F. (2007) Radon measurements in the SE and NE flank of Mt. Etna (Italy). *Radiation Measurements,* Vol. 42, 1404-1408

La Delfa, S., Agostino, I., Morelli, D., Patanè, G. (2008). Soil Radon concentration and effective stress variation at Mt Etna (Sicily) in the period January 2003-April 2005 *Radiation Measurements* 43 1299- 1304.

Martinelli, G. (1992). Fluidodynamical and chemical features of radon 222 related to total gases: implications on earthquakes prediction topics. IAEA-TECDOC-726 Isotopic and geochemical precursors of earthquakes and volcanic eruptions *Proceedings of an Advisory Group Meeting held in Vienna, 9-12 September 1991,* 48-62.

Morelli, D., Immè, G., La Delfa, S. , Lo Nigro, S., Patanè, G. (2006). Evidence of soil Radon as tracer of magma uprising at Mt. Etna. *Radiation Measurements* Vol. 41, 721-725

Morelli, D., Immè, G., Altamore, I., Cammisa, S. , Giammanco, S. , La Delfa, S. , Mangano, G., Neri, M., Patanè, G. (2011). Radionuclide measurements, via different methodologies, as tool for geophysical studies on Mt. Etna. *Nuclear Instruments and Methods in Physics Research A,* DOI NIMA 10.1016/j.nima.2011.01.172

Patanè, G., Coco, G., Corrao, M., Imposa, S. Montalto, A. (1995): Source parameters of seismic events at Mount Etna Volcano, Italy, during the outburst of the 1991-1993 eruption. *Phys. earth and Planet. Inter.,* 89, 149-162

Planinic, J., Radolic, V., Lazanin, Z. (2001). Temporal variation of radon in soil related to earthquakes. *Applied Radiation and Isotopes* ,Vol. 55, 267-272.

Ramola, R.C., Sing, S. and Virk, H.S. (1988). The correlation between radon anomalies and magnitude of earthquakes. *Nucl. Tracks Radiat. Meas.* ,Vol. 15, 689-692.

Rastogi, B.K., Chadha, R.K., Raju, I.P., (1986). Seismicity near Bhatsa reservoir, Maharashtra, India. *Physics of the Earth and Planetary Interiors* 44 (2), 179–199.

Rikitake, T., (1976). Earthquake prediction developments in solid earth. *Geophysics.,* Vol.9 357

Sabol, J. and Weng, P.-S. (1995). *Introduction to radiation protection dosimetry.* World Scientific, Singapore

Scholz, C.H., Sykes, L.R. and Aggarwal, Y. P.(1973). Earthquake prediction a physical basis. *Science,* Vol. 181, 803-810.

Shiratoi, K., (1927). The variation of radon activity of hot spring . Science Reports of the Tohoku Imperial University, Series 3 (16), 1725-1730.

Spitsyn, V. I., (1926). Collection of studies on radium and radioactive ores, 264

Tanner, A.B. (1959). Meterological influence on radon concentration in drill holes. *Mining Engineering,* Vol.11, 706-708.

Tanner, A.B. (1964). Radon migration in the ground : a supplementary review. The natural radiation environment. In Lowder, W.M. (Ed.). symposium proc. Houston, Texas, April 10-13, 1963. University of Chicago Press. Chicago, III, 161-190.

Singh, M., Ramola, R. C., Singh, B. , Singh S. and Virk, H.S.(1991). Radon anomalies: correlation with seismic activities in northern India. *Proceedings of the Second workshop on Radon Monitoring in Radioprotection, Environmental and/or Earth Science,* Furlan, G. and Tommasino, L. (Ed.) World Scientific.,354-375.

Ulomov, V.I., Zakharovc, A, I. and Ulomova, N.V. (1967). Tashkent earthquake of April 26, 1966, and its aftershocks. *Akad Nauk SSSR, Geophysic* 177, 567-570

Virk, H. S. and Singh, B. (1994). Radon recording of Uttarkashi earthquakes. *Geophysical Research Letters,* Vol.21, No.8, pp.737-740

Wattananikorn, K., Kanaree, M. and Wiboolsake, S. (1998). Soil gas radon as an earthquake precursor: some considerations on data improvement. *Radiation Measurements* Vol. 29, No.6, pp.593-598.

Yalim, H.A., Sandikcioglu, A., Unal, R., Orhun, O., (2007). Measurements of radon concentrations in well waters near the Aksehir fault zone in Afyonkarahisar, Turkey. *Radiation Measurements* 42, 505–508.

9

Radon as an Earthquake Precursor – Methods for Detecting Anomalies

Asta Gregorič, Boris Zmazek, Sašo Džeroski,
Drago Torkar and Janja Vaupotič
Jožef Stefan Institute, Ljubljana
Slovenia

1. Introduction

Radon is one of many geophysical and geochemical phenomena that can be considered to be an earthquake precursor. Due to the non-linear dependence of earthquakes' initial conditions, the question about the predictability of earthquakes often arises (Geller, 1997). The successful prediction of earthquakes is yet to be accomplished, in terms of their magnitude, location and time, and much effort has been spent on this goal.

The term "earthquake precursor" is used to describe a wide variety of geophysical and geochemical phenomena that reportedly precede at least some earthquakes (Cicerone et al., 2009). The observation of these types of phenomena is one recent research activity which has aimed at reducing the effects of natural hazards. Among the different precursors, geochemistry has provided some high-quality signals, since fluid flows in the Earth's crust have a widely recognised role in faulting processes (Hickman et al., 1995). The potential of gas geochemistry in seismo-tectonics has been widely discussed by Toutain and Baubron (1999).

In the late 1960s and early 1970s, reports from seismically active countries such as the former USSR, China, Japan and the USA (Ulomov & Mavashev, 1967; Wakita et al., 1980) indicated that concentrations of radon gas in the earth apparently changed prior to the occurrence of nearby earthquakes (Lomnitz, 1994). The noble gas radon (^{222}Rn) originates from the radioactive transformation of ^{226}Ra in the ^{238}U decay chain in the Earth's crust. Since radon is a radioactive gas, it is easy and relatively inexpensive to monitor instrumentally, and its short half-life (3.82 days) means that short-term changes in radon concentration in the earth can be monitored with a very good time resolution. Radon emanation from grains depends mainly on their ^{226}Ra content and their mineral grain size, its transport in the earth being governed by geophysical and geochemical parameters (Etiope & Martinelli, 2002), while exhalation is controlled by hydrometeorological conditions. The stress-strain developed within the Earth's crust before an earthquake leads to changes in gas transport and a rise of volatiles from the deep earth up to the surface (Ghosh et al., 2009; Thomas, 1988), resulting in anomalous changes in radon concentration. The mechanism of observed radon anomalies is still poorly understood, although several theories have been proposed (Atkinson, 1980; King, 1978; Lay et al., 1998; Martinelli, 1991). Over the past three decades, the occurrence of anomalous temporal changes of radon concentrations has been studied by several authors specialising in soil gas (King, 1984, 1985; Kuo et al., 2010; Mogro-Campero et al., 1980; Planinić et al., 2001; Ramola et al., 2008; Ramola et al., 1990; Reddy & Nagabhushanam, 2011; Walia et al., 2009a; Walia et al., 2009b; Yang et al., 2005; Zmazek et al., 2005; Zmazek et al., 2002b) and groundwater (Barragán et al., 2008; Favara et al., 2001; Gregorič et al., 2008; Heinicke et al., 2010; Kuo et al., 2006; Ramola, 2010; Singh et al., 1999; Zmazek et al., 2002a; Zmazek et al., 2006). However, radon anomalies are not only controlled by seismic activity but also by meteorological parameters like soil moisture, rainfall, temperature and barometric pressure (Ghosh et al., 2009; Stranden et al., 1984). This makes it complicated and, for small earthquakes, often impossible to differentiate between those anomalies caused by seismic events and those caused solely by atmospheric changes. Therefore, the application of theoretical and empirical algorithms for removing meteorological effects is necessary (Choubey et al., 2009; Ramola et al., 2008; Ramola et al., 1988; Torkar et al., 2010; Zmazek et al., 2003). In this chapter, the different approaches to distinguishing between those anomalies in radon time series caused by seismic activity and those caused solely by hydrometeorological parameters are presented and discussed.

2. Radon migration in the Earth's crust

Only a fraction of the radon atoms created by radioactive transformation from radium are able to emanate from mineral grains and enter into the void space, filled either by gas or water. Radon ascends towards the surface mainly through cracks or faults, on a short scale by diffusion and, for longer distances, by advection - dissolved either in water or in carrier gases. Gas movement should be ascribed to the combination of both processes. Diffusive movement is driven by a concentration gradient and is described by Fick's law. Considering gas diffusion in porous media, it is necessary to take into account that the volume through which gas diffuses is reduced and the average path length between two points is increased, thus altering the diffusion coefficient (Etiope & Martinelli, 2002). Nevertheless, the velocity of radon transport in the earth is quite low ($\leq 10^{-3}$ cm/s) and the concentration of radon is reduced by radioactive decay to the background level before even 10 m are traversed (Etiope & Martinelli, 2002; Fleischer, 1981). Diffusion is only important in capillaries and small-pored rocks. On the other hand, the velocity and space scales of advective movements are much higher than those of diffusive ones. Advective transport is driven by pressure gradients, following Darcy's law. The amount of radon itself is, however, too small to form a macroscopic quantity of gas which can react to pressure gradients. Therefore, it must be carried by a macroscopic flow of carrier gases (Kristiansson & Malmqvist, 1982). A gas mixture formed by carrier gases (e.g., CO_2, CH_4, and N_2) and rare gases (e.g., He, Rn) can be referred to as "geogas" (Etiope & Martinelli, 2002; Kristiansson & Malmqvist, 1982). In dry, porous or fractured media, gas flows through an interstitial or fissure space (gas-phase advection) whereas in saturated, porous media gas can dissolve and then be transported in three ways: by groundwater (water-phase advection), by displacing water (gas-phase advection) or by forming a bubble flow by means of buoyancy in aquifers and water-filled fractures. The bubble movement has been theoretically and experimentally recognised as a fast gas migration mechanism governing the distribution of carrier and trace gases over wide areas on the Earth's surface (Vàrhegyi et al., 1992).

2.1 External effects on radon in soil gas and water

Radon concentration in soil, gas or water is not only controlled by geophysical parameters, but it also changes due to other external effects. Meteorological effects – such as soil humidity, rainfall, temperature, barometric pressure and wind – control radon concentrations in soil gas. These parameters change the physical characteristics of soil and rock, thus influencing the rate of radon transport and, consequently, perturbing eventual radon variations caused by geophysical processes originating in the deeper parts of the Earth's crust. Shallow soil levels are more affected by changing meteorological conditions than deeper ones. Radon concentrations with no larger variations present are usually observed at depths of 0.8 m or deeper. Besides the effects of meteorological parameters on radon in soil gas, considerable variations in the gas composition of thermal springs have been shown to be the result of fluctuations of local hydrologic regimes (Klusman & Webster, 1981).

The significant influence of barometric pressure has been discussed by several authors, who clearly pointed out an inverse relationship between barometric pressure and radon concentration in soil gas (Chen et al., 1995; Clements & Wilkening, 1974; Klusman & Webster, 1981). A decrease in barometric pressure, with the values of other environmental parameters remaining constant, generally causes an increase in radon exhalation from the ground, whereas during periods of rising pressure, air with low radon concentration is forced into the ground, thus diluting radon. Temperature-related fluctuations of soil gas radon concentration have also been proven to be very important. Klusman & Jaacks (1987) found an inverse relationship between soil temperature and radon concentration. They suggested that lower air temperatures as compared with soil temperatures during winter months promoted an upward movement of radon by convection, whereas during the summer, lower soil temperatures as compared with air temperatures and an inversion layer below the level of sampling reduces the upward flux and observed concentration. In general, the behaviour of soil gas migration in different types of soil is seasonally dependent (King & Minissale, 1994; Washington & Rose, 1990). In systems where gas movement is driven by diffusion or slow advection processes, radon activity in soil might be controlled by soil moisture and rainfall through the opening of cracks in the surface (Pinault & Baubron, 1996; Toutain & Baubron, 1999). On the other hand, barometric pressure has the major influence on radon concentrations in soils in advective systems, which display generally higher gas flows. However, micro-scale soil heterogeneities in permeability, porosity and lithology can cause significant heterogeneities in the response of radon concentration to changes of atmospheric parameters (King & Minissale, 1994; Neznal et al., 2004). Numerous and often divergent results in studies related to the effect of external factors on soil gas radon concentration suggest that no general predictive model for excluding meteorological effects can be proposed, and studies of radon in soil gas need a simultaneous record of meteorological parameters.

3. Anomalous radon concentration and seismicity

Both mechanisms of radon transport – diffusion and advection – depend on both soil porosity and permeability, which at the same time vary as a function of the stress field (Holub & Brady, 1981). However, migration by diffusion is negligible, where a component of advective long-distance transport exists (Etiope & Martinelli, 2002). The high permeability of the bedrock and soil in areas of crustal discontinuities, such as fractures and fault zones, promotes intense degassing fluxes, which causes higher soil gas radon concentrations on the ground surface above active fault zones. Although several measurements, experiments and models have been performed, the understanding of the mechanism of radon anomalies and their connection to earthquakes is still inadequate (Chyi et al., 2010; King, 1978; Ramola et al., 1990). Before the earthquake stress in the Earth's crust builds up causing a change in the strain field; the formation of new cracks and pathways under the tectonic stress leads to changes in gas transport and a rise in volatiles from the deep layers to the surface. In fact, fluids play a widely recognised role in controlling the strength of crustal fault zones (Hickman et al., 1995). Anomalous changes of radon concentration are closely linked to changes in fluid flow and, therefore, also to highly permeable areas along fault zones. Large faults are not discrete surfaces but rather a braided array of slip surfaces encased in a highly fractured and often hydrothermally altered transition – or "damage" – zone. Episodic fracturing and brecciation are followed by cementation and crack healing, leading to cycles of permeability enhancement and reduction along faults (Hickman et al., 1995).

Several mechanisms have been proposed, which could explain the relationship between radon anomalies and earthquake. Two models of earthquake precursors are discussed by Mjachkin et al. (1975), with a common principle: at a certain preparation stage, a region of many cracks is formed. According to the dilatancy-diffusion model (Martinelli, 1991; Mjachkin et al., 1975), the increase in tectonic stress causes the extension and opening of favourably-oriented cracks in a porous, cracked, saturated rock. Water flows into the opened cracks, drying the rock near each pore and finally resulting in a decrease of pore pressure in the total earthquake preparation zone. Water from the surrounding medium diffuses into the zone. The increased water-rock surface area, due to cracking, leads to an increase in radon transfer from the rock matrix to the water. At the end of the diffusion period, the appearance of pore pressure and the increased number of cracks leads to the main rupture. According to the crack-avalanche model (Mjachkin et al., 1975), the increasing tectonic stress leads to the formation of a cracked focal rock zone, with slowly altering volume and shape. At a certain stage – when the whole focal zone becomes unstable – the cracks quickly concentrate near the fault surface, triggering the main rupture. An alternate mechanism for earthquake precursory study, based on stress-corrosion theory, has been proposed by Anderson and Grew (1977). According to them, the observed radon anomalies are due to slow crack growth controlled by stress corrosion in a rock matrix saturated by ground waters. King (1978) has proposed a compression mechanism for radon release, whereby anomalous high radon release may be due to an increase of crustal compression before an impending earthquake, that squeezes out soil gas into the atmosphere at an increasing rate.

Toutain and Baubron (1999) observed that gas transfer within the upper crust is affected by strains less than 10^{-7}, much smaller than those causing earthquakes. According to Dobrovolsky (1979), the radius of the effective precursory manifestation zone depends on the earthquake magnitude and can be calculated using the empirical equation:

$$R_D = 10^{0.43 \times M_L} \tag{1}$$

Where R_D is the strain radius in km and M_L is the magnitude of the earthquake. Considering the Earth's crust to be an anisotropic medium, this law can be modified according to the effective sensitivity to the impending earthquake. The ideal circle with its theoretical radius can be transformed into an ellipse or characterised by shadow areas where no precursory phenomena are observable due to crustal anisotropy, discontinuities or loose contacts along some faults, which prevent further stress transfer (İnan & Seyis, 2010; Martinelli, 1991).

Although radon anomalies can be studied in soil gas and thermal waters, thermal waters could be much more representative of the geologic environment and could be more reactive to stress/strain changes acting at depth than soil gases. The disadvantage of soil gases lie in weak gas concentrations, generally due to the thickness of the sedimentary cover and the high level of atmospheric perturbations (Toutain & Baubron, 1999).

4. Methods for detecting anomalies in radon time series

An anomaly in radon concentration is defined as a significant deviation from the mean value. Due to the high background noise of radon time series, it is often impossible to distinguish an anomaly caused solely by a seismic event from one resulting from meteorological or hydrological parameters. For this reason, the implementation of more advanced statistical methods in data evaluation is important (Belyaev, 2001; Cuomo et al., 2000; Negarestani et al., 2003; Sikder & Munakata, 2009; Steinitz et al., 2003). In our research, radon has been monitored in several thermal springs (Gregorič et al., 2008; Zmazek et al., 2002a; Zmazek et al., 2006) and in soil gas (Zmazek et al., 2002b) and different approaches to distinguishing radon anomalies were applied.

4.1 Standard deviation

A very common practice in determining radon anomalies is the use of standard deviation. The average radon concentration is calculated for different periods with regard to the nature of yearly cycles of radon concentration. In the case of radon in soil gas, the mean value of radon concentration is calculated separately for four seasons (spring, summer, autumn and winter) based on the air and soil temperature.

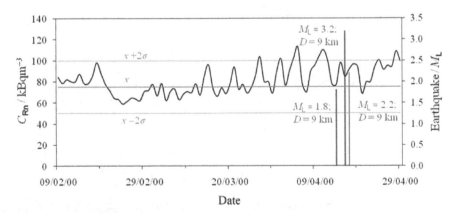

Fig. 1. Continuous radon concentration recorded in soil gas at Krško basin. Straight lines represent the mean value and two standard deviations of the radon concentration. Local seismicity is expressed in terms of local magnitude (M_L) and the distance between the measuring location and the earthquake epicentre (D). Radon anomalies are C_{Rn} values outside the ±2σ region.

In contrast to soil gas, radon in ground or spring water is greatly influenced by the hydrologic cycle, which has to be considered during the data analysis. To define the mean and standard deviation, anomalously high and low values – which may cause unnecessary high deviation and perturb the real anomalies – have to be neglected. The periods when radon concentration deviates by more than ±2σ from the related seasonal value are considered as radon anomalies that are possibly caused by earthquake events and not by meteorological parameters (Ghosh et al., 2007; Gregorič et al., 2008; Virk et al., 2002; Zmazek et al., 2002b). Fig. 1 shows an anomalous radon concentration, exceeding 2σ above the average value, which appeared approximately 10 days before the occurrence of three earthquakes with magnitudes from 1.8 to 3.2.

4.2 Relationship between radon exhalation and barometric pressure

An inverse relationship exists between the time derivative of radon concentration in soil gas and the time derivative of barometric pressure (as was discussed previously in section 2.1). A decrease in barometric pressure causes an increase in radon exhalation from the ground, whereas during periods of rising pressure, air with low radon concentration is forced into

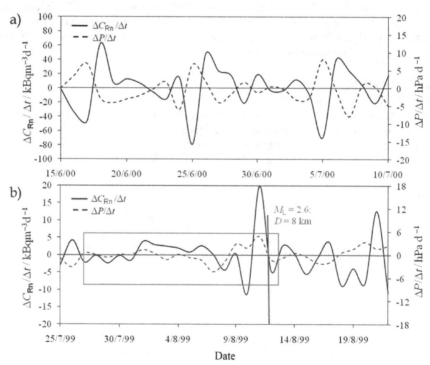

Fig. 2. The time gradient of radon concentration in soil gas and the time gradient of barometric pressure during two periods at the Krško basin: a) the period without local seismic activity, b) the seismically active period, whereby the radon anomaly 14 days before the earthquake is marked by the green rectangle. The earthquake is expressed in terms of local magnitude (M_L) and the distance between the measuring location and the earthquake epicentre (D).

the ground, thus diluting the radon concentration. Therefore, deviations from this rule during these periods – when the time gradient of barometric pressure, $\Delta P/\Delta t$, and the time gradient of radon concentration, $\Delta C_{Rn}/\Delta t$, in soil gas have the same sign – can be considered to be radon anomalies (Zmazek et al., 2002b). A clear negative correlation between the time gradient of radon concentration and the time gradient of barometric pressure can be seen in Fig. 2a, when no seismic activity is present. The radon anomaly, characterised by positive correlation of time gradients, is marked in Fig. 2b. Anomalous behaviour in radon concentration started 14 days before the earthquake with a local magnitude of 2.6, and ended a few days after the earthquake.

4.3 Machine learning methods

Machine learning methods have been successfully applied to many problems in the environmental sciences (Džeroski, 2002). In the case of radon as an earthquake precursor, it must be considered – as discussed in section 2.1 – that the variation in radon concentration is controlled not only by geophysical phenomena in the Earth's crust, but also by the environmental parameters associated with the radon monitoring sites. With machine learning methods, a model for the prediction of radon concentration can be built, taking into account various environmental parameters (e.g., barometric pressure, rainfall, and air and soil temperature). The aim is to identify radon anomalies which might be caused by seismic events. The application of artificial neural networks (Negarestani et al., 2002, 2003; Torkar et al., 2010), regression and model trees (Džeroski et al., 2003; Sikder & Munakata, 2009; Zmazek et al., 2003; Zmazek et al., 2006) and some other methods (Sikder & Munakata, 2009; Steinitz et al., 2003) have proven to be useful means of extracting radon anomalies caused by seismic events.

4.3.1 Artificial neural networks

An artificial neural network (ANN) is a well-known computational structure inspired by the operation of the biological neural system (Jain et al., 1996) and it is a well-established tool, being used widely in signal processing, pattern recognition and other applications. An ANN consists of a set of units (neurons, nodes), and a set of weighted interconnections among them (links). The organisation of neurons and their interconnections defines the net topology. The inputs are grouped in an input layer, the outputs in an output layer and all the other units in so-called hidden layers. The algorithm repeatedly adjusts the weights to minimise the mean square error between the actual output vector and the desired network output vector. The universal approximator functional form of ANNs is well-suited for the requirements of modelling the non-linear dependency of radon concentrations on multiple variables. Among a number of various topologies, training algorithms and architectures of ANNs, the traditional multilayer perceptron (MLP) with a conjugate gradient learning algorithm was chosen in the case of analysing the soil gas radon concentration time series at the Krško basin (Torkar et al., 2010). The series was first split into seismically non-active periods (NSA) and seismically active periods (SA), adjusting the duration of the seismic window from 0 to 10 days before and after the earthquake and with the purpose of investigating the influence of a complete earthquake event on radon concentration (the preparation phase, the earthquake itself and aftershocks). The ANN of the MLP type was trained with each of the NSA datasets, which were divided into three sets: the training set (60%), the cross-validation set (15%) and the test set (25%). The ANN was trained with the training and cross-validation set, while the test set was used to verify its performance. The topology of the ANN generated for each NSA dataset is shown in Fig. 3.

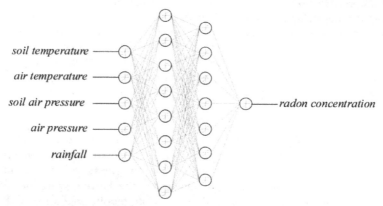

Fig. 3. The ANN topology for learning radon concentration dependency on environmental parameters.

In the testing phase, the correlation between the measured (m-C_{Rn}) and predicted (p-C_{Rn}) radon concentration in NSA periods was compared to the correlation between the measured and predicted radon concentration in the entire dataset (NSA and SA). The difference between the correlation coefficients might indicate a period of seismically induced radon anomaly. The ratio between the measured and predicted values (m-C_{Rn}/p-C_{Rn})–1 represents the discrepancy between both values (Fig. 4).

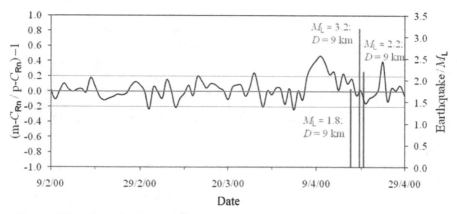

Fig. 4. The ratio between the measured and predicted radon concentration (m-C_{Rn}/p-C_{Rn})–1 using an ANN in the case of soil gas radon in the Krško basin for a seismic window of ±7 days. A radon anomaly, possibly caused by a seismic event, is observed when the signal exceeds the threshold value of 0.2.

A radon anomaly is held to be when the absolute value of signal (m-C_{Rn}/p-C_{Rn})–1 exceeds the predefined threshold of 0.2. The ANN in this case performed the best in the case of a seismic window of ±7 days (indicating the length of the period of pre- and post-seismic changes).

4.3.2 Decision trees

Decision trees are machine-learning methods for constructing prediction models from data. The models are obtained by recursively partitioning the data space and fitting a simple prediction model within each partition. As a result, the partitioning can be represented graphically as a decision tree, where each internal node contains a test on an attribute, each branch corresponds to an outcome of the test, and each leaf node gives a prediction for the value of the class variable (Džeroski, 2001; Loh, 2011). Regression trees are designed for dependent variables that take continuous or ordered discrete values. Like classical regression equations, they predict the value of a dependent variable (called the class) from the values of a set of independent variables (called attributes).

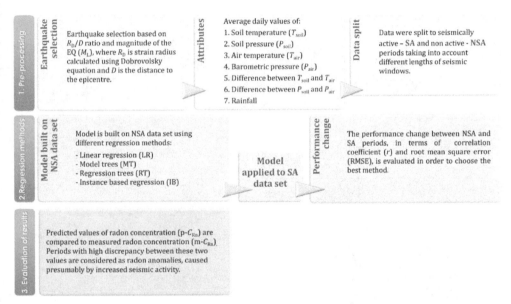

Fig. 5. A schematic description of the different stages of radon data series analysis with machine learning methods.

The model in each leaf can be either a linear equation or just a constant; trees with linear equations in the leaves are also called model trees. Tree construction proceeds recursively, starting with the entire set of training examples. At each step, the most discriminating attribute is selected as the root of the sub-tree and the current training set is split into subsets according to the values of the selected attribute. For continuous attributes, a threshold is selected and two branches are created, based on that threshold. The attributes that appear in the training set are considered to be thresholds. Tree construction stops when the variance of the class values of all examples in a node is small enough. These nodes are called leaves and are labelled with a model for predicting the class value. An important mechanism used to prevent the tree from over-fitting data is tree pruning.

Regression (RT) and model trees (MT), as implemented with the WEKA data mining suite (Witten & Frank, 1999), were used for predicting radon concentration from meteorological parameters in the case of radon time series in soil gas at the Krško basin (Zmazek et al., 2003; Zmazek et al., 2005) and in the thermal spring water in Zatolmin (Zmazek et al., 2006).

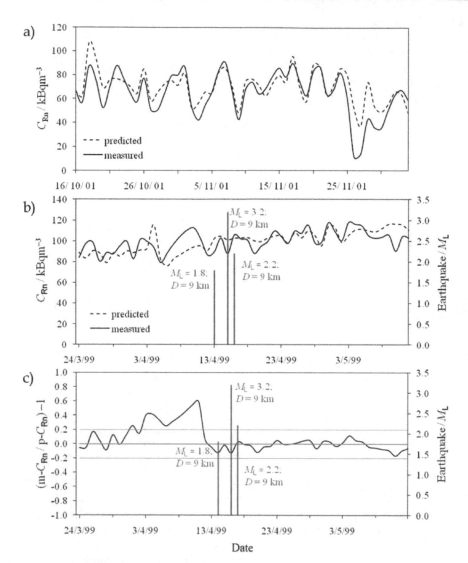

Fig. 6. Measured and predicted radon concentration using model trees in the case of soil gas radon at the Krško basin for a seismic window ±7 days; a) low discrepancy in the period without seismic activity; b) high discrepancy starting 10 days before a group of earthquakes; c) the ratio between the measured and predicted radon concentration $(m-C_{Rn}/p-C_{Rn})-1$ for the same SA period. A radon anomaly, possibly caused by earthquakes, is observed when the signal exceeds the threshold value of 0.2 (marked by the green lines).

As presented in Fig. 5 the first stage of data analysis comprises the selection of attributes – i.e. environmental parameters – and the partitioning of the whole data set to the periods with and without seismic activity, SA and NSA respectively. After inspecting the correlation changes between radon concentration and barometric pressure, a seismic window of ±7 days was chosen. The performance was estimated with 10-fold cross-validation in order to evaluate the predictability of the radon concentration in the NSA periods. The model built on the NSA data set was then applied to the SA data set and the performance change was determined using two different measures, the correlation coefficient (r) and the root mean square error (RMSE). For the purposes of prediction, the measured performance in NSA periods should be higher than the performance in SA periods. In these periods, when the discrepancy between the measured and predicted radon concentration is low, no seismic activity is anticipated (Fig. 6a), while in the periods with a higher discrepancy, a radon anomaly can be ascribed to increased seismic activity, rather than to the effect of atmospheric parameters (Fig. 6b). This discrepancy is clearly shown in form of the ratio between both values (m-C_{Rn}/p-C_{Rn})–1, as shown in Fig. 6c. A radon anomaly is held to be when the absolute value of the signal (m-C_{Rn}/p-C_{Rn})–1 exceeds the predefined threshold of 0.2. Besides regression trees, other machine learning methods were also tested (e.g., linear regression and instance-based regression). However, model trees have been shown to outperform other approaches.

4.4 Comparison of the results

The results of all of the approaches used for the identification of radon anomalies caused by seismic events in the case of soil gas radon at the Krško basin are shown in Fig. 7 for the period of 1/9 – 30/12/2000. Among all of the approaches – and although not very exact – the ±xσ method (I) is the most frequently used. The threshold of anomalous concentrations (e.g., ±1σ, ±2σ, ±3σ) should be chosen in order to minimise the number of false anomalies (FA: anomalies in seismically non-active periods) and so as not to miss the correct ones (CA). Generally, a range of ±2σ from the related seasonal mean value is chosen. Furthermore, a cyclic behaviour of radon concentration has to be taken into account in order to accurately define the period of standard deviation and the calculation of the mean value. For this purpose different methods of time series analyses – for example, Fourier transform (Ramola, 2010) – can be applied.

In the case shown in Fig. 7a, three radon anomalies exceeding 2σ above the mean value may be noticed. The first, in the beginning of September, cannot be assigned to a seismic event (FA). About a week before a weak earthquake of local magnitude M_L=1.1, 5 km away from the measurement location – which is the first of five earthquakes over a period of 2 months – the second anomaly is observed. And finally, the third one can be noticed soon after a weak earthquake 6 km away (M_L=1).

The first of the anomalies mentioned above as FA is also visible by applying the method of pressure gradients (II) (Fig. 7b). A positive correlation between the time gradient of radon concentration and the time gradient of barometric pressure is considered to be a radon anomaly, and corresponds to the anomaly observed through method (I) which preceded the first earthquake (M_L=1.1). A radon anomaly can also be noticed a few days before the last earthquake, as with the analysis of method (I). Additionally, the anomalous behaviour of the radon concentration as regards the gradient approach is observed during the period starting a few days before the earthquake with M_L=2.7 and lasting until the earthquake with M_L=1.

More often than not, swarms of anomalies are observed over longer periods, with a higher number of anomalies in a swarm observed for approach (II) than for approach (I). As an additional criterion, a threshold of $\Delta P/\Delta t > 2$ hPa d^{-1} is introduced by this approach in order to optimise the identification of anomalies caused by seismic events. However, by increasing the threshold value above 2 hPa d^{-1}, the ratio between correct and false anomalies cannot be significantly improved (Zmazek et al., 2005).

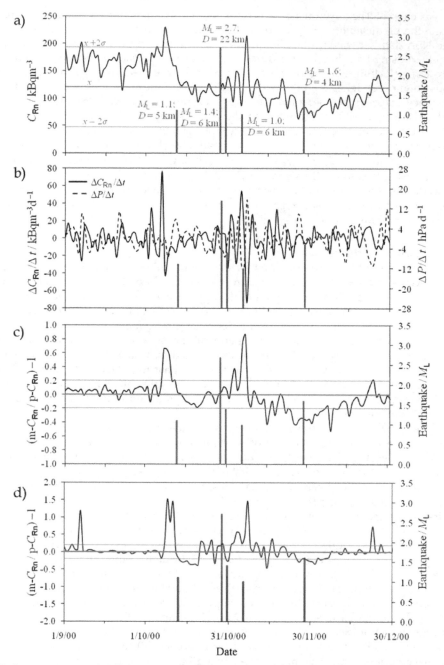

Fig. 7. A comparison of different approaches for the identification of radon anomalies: a) standard deviation (I); b) the relationship between radon exhalation and barometric pressure (II); c) artificial neural networks (III); and d) model trees (IV).

Both machine learning approaches, artificial neural networks (III) and decision trees (IV) give promising results, with a low number of false anomalies. The two distinctive anomalies – observed in Fig. 7c and Fig. 7d, for ANN and MT, respectively – confirm the anomalies identified by approaches (I) and (II). Additionally, a relatively long negative anomaly was observed using the ANN approach at the end of November, accompanying the earthquake with M_L=1.6. On the other hand, the same negative anomaly is only weakly expressed using the MT approach. A FA observed at the beginning of September using approaches (I) and (II) was also noticed using the MT approach but not by the ANN approach. Approaches (III) and (IV) do not appear to greatly depend upon the choice for the threshold of $(m\text{-}C_{Rn}/p\text{-}C_{Rn})\text{--}1$ and can, therefore, be used with less hesitation.

5. Conclusion

Since the appropriate interpretation of field measurements plays an important role in any research, the purpose of this work was to combine and evaluate the different approaches applied by our research group for differentiating the radon anomalies caused by increased seismic activity from those caused solely by environmental parameters. The application of four different approaches – standard deviation from the related mean value (I), the correlation between time gradients of barometric pressure and radon concentration (II), artificial neural networks (III) and decision trees (IV) – was presented. Radon anomalies based on approach (I) have been less successful in predicting earthquakes than those based on the other three approaches. Secondly, approaches (I) and (II) greatly depend upon the values of the $\pm x\sigma$ and $\Delta P/\Delta t$ thresholds, respectively, while the dependence of approaches (III) and (IV) on the threshold of $(m\text{-}C_{Rn}/p\text{-}C_{Rn})\text{--}1$ is very weak. The number of false anomalies for approach (II) points to the disturbance of radon exhalation by other environmental parameters and not just by barometric pressure. The assumption that radon exhalation is only directly influenced by barometric pressure is further suggested by different forms of radon transport at compression and dilatation zones (Ghosh et al., 2009). Promising results are achieved by applying approaches (III) and (IV), which make it possible to simultaneously incorporate all of the available environmental parameters. Furthermore, in using these techniques, the relation between radon concentration and environmental parameters does not necessarily have to be presumed linear. And finally, in taking into account the scale of the earthquake magnitudes observed during the time of radon measurements, one may speculate that the performance of the applied approaches would be better in the case of stronger earthquakes.

6. Acknowledgement

This study was done within the program P1-0143: Cycling of substances in the environment, mass balances, modelling of environmental processes and risk assessment.

7. References

Anderson, O.L., & Grew, P.C. (1977). Stress-corrosion theory of crack-propagation with applications to geophysics. *Reviews of Geophysics*, Vol. 15, No. 1, pp. 77–104

Atkinson, B.K. (1980). Stress corrosion and the rate-dependent tensile failure of a fine-grained quartz rock. *Tectonophysics*, Vol. 65, No. 3–4, pp. 281–290

Barragán, R.M., Arellano, V.M., Portugal, E., & Segovia, N. (2008). Effects of changes in reservoir thermodynamic conditions on ^{222}Rn composition of discharged fluids: study for two wells at Los Azufres geothermal field (Mexico). *Geofluids*, Vol. 8, No. 4, pp. 252-262

Belyaev, A.A. (2001). Specific features of radon earthquake precursors. *Geochemistry International*, Vol. 39, No. 12, pp. 1245-1250

Chen, C., Thomas, D.M., & Green, R.E. (1995). Modeling of radon transport in unsaturated soil. *Journal of Geophysical Research-Solid Earth*, Vol. 100, No. B8, pp. 15517-15525

Choubey, V.M., Kumar, N., & Arora, B.R. (2009). Precursory signatures in the radon and geohydrological borehole data for M4.9 Kharsali earthquake of Garhwal Himalaya. *Science of the Total Environment*, Vol. 407, No. 22, pp. 5877-5883

Chyi, L.L., Quick, T.J., Yang, T.F., & Chen, C.H. (2010). The experimental investigation of soil gas radon migration mechanisms and its implication in earthquake forecast. *Geofluids*, Vol. 10, No. 4, pp. 556-563

Cicerone, R.D., Ebel, J.E., & Britton, J. (2009). A systematic compilation of earthquake precursors. *Tectonophysics*, Vol. 476, No. 3-4, pp. 371-396

Clements, W.E., & Wilkening, M.H. (1974). Atmospheric-pressure effects on Rn-222 transport across earth-air interface. *Journal of Geophysical Research*, Vol. 79, No. 33, pp. 5025-5029

Cuomo, V., Di Bello, G., Lapenna, V., Piscitelli, S., Telesca, L., Macchiato, M., & Serio, C. (2000). Robust statistical methods to discriminate extreme events in geoelectrical precursory signals: Implications with earthquake prediction. *Natural Hazards*, Vol. 21, No. 2-3, pp. 247-261

Dobrovolsky, I.P., Zubkov, S.I., & Miachkin, V.I. (1979). Estimation of the size of earthquake preparation zones. *Pure and Applied Geophysics*, Vol. 117, No. 5, pp. 1025-1044

Džeroski, S. (2001). Data mining in a nutshell, In: *Relational data mining*, Džeroski, S., Lavrač, N. (Eds.), pp. 3-27, Springer, Berlin

Džeroski, S. (2002). Environmental sciences, In: *Handbook of data mining and knowledge discovery*, Klösgen, W., Żytkow, J. (Eds.), pp. 817-830, Oxford University Press, Oxford

Džeroski, S., Todorovski, L., Zmazek, B., Vaupotič, J., & Kobal, I. (2003). Modelling soil radon concentration for earthquake prediction. *Discovery Science, Proceedings*, Vol. 2843, pp. 87-99

Etiope, G., & Martinelli, G. (2002). Migration of carrier and trace gases in the geosphere: an overview. *Physics of the Earth and Planetary Interiors*, Vol. 129, No. 3-4, pp. 185-204

Favara, R., Grassa, F., Inguaggiato, S., & Valenza, M. (2001). Hydrogeochemistry and stable isotopes of thermal springs: earthquake-related chemical changes along Belice Fault (Western Sicily). *Applied Geochemistry*, Vol. 16, No. 1, pp. 1-17

Fleischer, R.L. (1981). Dislocation model for radon response to distant earthquakes. *Geophysical Research Letters*, Vol. 8, No. 5, pp. 477-480

Geller, R.J. (1997). Earthquake prediction: a critical review. *Geophysical Journal International*, Vol. 131, No. 3, pp. 425-450

Ghosh, D., Deb, A., & Sengupta, R. (2009). Anomalous radon emission as precursor of earthquake. *Journal of Applied Geophysics*, Vol. 69, No. 2, pp. 67-81

Ghosh, D., Deb, A., Sengupta, R., Patra, K.K., & Bera, S. (2007). Pronounced soil-radon anomaly - Precursor of recent earthquakes in India. *Radiation Measurements*, Vol. 42, No. 3, pp. 466–471

Gregorič, A., Zmazek, B., & Vaupotič, J. (2008). Radon concentration in thermal water as an indicator of seismic activity. *Collegium Antropologicum*, Vol. 32, pp. 95–98

Heinicke, J., Italiano, F., Koch, U., Martinelli, G., & Telesca, L. (2010). Anomalous fluid emission of a deep borehole in a seismically active area of Northern Apennines (Italy). *Applied Geochemistry*, Vol. 25, No. 4, pp. 555–571

Hickman, S., Sibson, R., & Bruhn, R. (1995). Introduction to special section: Mechanical involvement of fluids in faulting. *Journal of Geophysical Research*, Vol. 100, No. B7, pp. 12831–12840

Holub, R.F., & Brady, B.T. (1981). The effect of stress on radon emanation from rock. *Journal of Geophysical Research*, Vol. 86, No. NB3, pp. 1776–1784

İnan, S., & Seyis, C. (2010). Soil radon observations as possible earthquake precursors in Turkey. *Acta Geophysica*, Vol. 58, No. 5, pp. 828–837

Jain, A.K., Jianchang, M., & Mohiuddin, K.M. (1996). Artificial Neural Networks: A Tutorial. *Computer*, Vol. 29, No. 3, pp. 31–44

King, C.Y. (1978). Radon emanation on San Andreas Fault. *Nature*, Vol. 271, No. 5645, pp. 516–519

King, C.Y. (1984). Impulsive radon emanation on a creeping segment of the San-Andreas fault, California. *Pure and Applied Geophysics*, Vol. 122, No. 2–4, pp. 340–352

King, C.Y. (1985). Radon monitoring for earthquake prediction in China. *Earthquake Prediction Research*, Vol. 3, No. 1, pp. 47–68

King, C.Y., & Minissale, A. (1994). Seasonal variability of soil-gas radon concentration in central California. *Radiation Measurements*, Vol. 23, No. 4, pp. 683–692

Klusman, R.W., & Jaacks, J.A. (1987). Environmental influences upon mercury, radon and helium concentrations in soil gases at a site near Denver, Colorado. *Journal of Geochemical Exploration*, Vol. 27, No. 3, pp. 259–280

Klusman, R.W., & Webster, J.D. (1981). Preliminary analysis of meteorological and seasonal influences on crustal gas emission relevant to earthquake prediction. *Bulletin of the Seismological Society of America*, Vol. 71, No. 1, pp. 211–222

Kristiansson, K., & Malmqvist, L. (1982). Evidence for nondiffusive transport of $^{222}_{86}Rn$ in the ground and a new physical model for the transport. *Geophysics*, Vol. 47, No. 10, pp. 1444–1452

Kuo, T., Fan, K., Kuochen, H., Han, Y., Chu, H., & Lee, Y. (2006). Anomalous decrease in groundwater radon before the Taiwan M6.8 Chengkung earthquake. *Journal of Environmental Radioactivity*, Vol. 88, No. 1, pp. 101–106

Kuo, T., Su, C., Chang, C., Lin, C., Cheng, W., Liang, H., Lewis, C., & Chiang, C. (2010). Application of recurrent radon precursors for forecasting large earthquakes (Mw > 6.0) near Antung, Taiwan. *Radiation Measurements*, Vol. 45, No. 9, pp. 1049–1054

Lay, T., Williams, Q., & Garnero, E.J. (1998). The core-mantle boundary layer and deep Earth dynamics. *Nature*, Vol. 392, pp. 461–468

Loh, W.-Y. (2011). Classification and regression trees. *Wiley Interdisciplinary Reviews: Data Mining and Knowledge Discovery*, Vol. 1, No. 1, pp. 14–23

Lomnitz, C. (1994). *Fundamentals of Earthquake Prediction*, John Wiley & Sons, New York

Martinelli, G. (1993). Fluidodynamical and chemical features of radon 222 related to total gases: implications for earthquake predictions, *Proceedings of IAEA Meeting on isotopic and geochemical precursors of earthquakes and volcanic eruptions,* Vienna, September 1991

Mjachkin, V.I., Brace, W.F., Sobolev, G.A., & Dieterich, J.H. (1975). Two models for earthquake forerunners. *Pure and Applied Geophysics,* Vol. 113, No. 1, pp. 169–181

Mogro-Campero, A., Fleischer, R.L., & Likes, R.S. (1980). Changes in subsurface radon concentration associated with earthquakes. *Journal of Geophysical Research,* Vol. 85, No. NB6, pp. 3053–3057

Negarestani, A., Setayeshi, S., Ghannadi-Maragheh, M., & Akashe, B. (2002). Layered neural networks based analysis of radon concentration and environmental parameters in earthquake prediction. *Journal of Environmental Radioactivity,* Vol. 62, No. 3, pp. 225–233

Negarestani, A., Setayeshi, S., Ghannadi-Maragheh, M., & Akashe, B. (2003). Estimation of the radon concentration in soil related to the environmental parameters by a modified Adaline neural network. *Applied Radiation and Isotopes,* Vol. 58, No. 2, pp. 269–273

Neznal, M., Matolin, M., Just, G., & Turek, K. (2004). Short-term temporal variations of soil gas radon concentration and comparison of measurement techniques. *Radiation Protection Dosimetry,* Vol. 108, No. 1, pp. 55–63

Pinault, J.L., & Baubron, J.C. (1996). Signal processing of soil gas radon, atmospheric pressure, moisture, and soil temperature data: A new approach for radon concentration modeling. *Journal of Geophysical Research-Solid Earth,* Vol. 101, No. B2, pp. 3157–3171

Planinić, J., Radolić, V., & Lazanin, Ž. (2001). Temporal variations of radon in soil related to earthquakes. *Applied Radiation and Isotopes,* Vol. 55, No. 2, pp. 267–272

Ramola, R.C. (2010). Relation between spring water radon anomalies and seismic activity in Garhwal Himalaya. *Acta Geophysica,* Vol. 58, No. 5, pp. 814–827

Ramola, R.C., Prasad, Y., Prasad, G., Kumar, S., & Choubey, V.M. (2008). Soil-gas radon as seismotectonic indicator in Garhwal Himalaya. *Applied Radiation and Isotopes,* Vol. 66, No. 10, pp. 1523–1530

Ramola, R.C., Singh, M., Sandhu, A.S., Singh, S., & Virk, H.S. (1990). The use of radon as an earthquake precursor. *Nuclear Geophysics,* Vol. 4, No. 2, pp. 275–287

Ramola, R.C., Singh, S., & Virk, H.S. (1988). A model for the correlation between radon anomalies and magnitude of earthquakes. *Nuclear Tracks and Radiation Measurements,* Vol. 15, No. 1–4, pp. 689–692

Reddy, D.V., & Nagabhushanam, P. (2011). Groundwater electrical conductivity and soil radon gas monitoring for earthquake precursory studies in Koyna, India. *Applied Geochemistry,* Vol. 26, No. 5, pp. 731–737

Sikder, I.U., & Munakata, T. (2009). Application of rough set and decision tree for characterization of premonitory factors of low seismic activity. *Expert Systems with Applications,* Vol. 36, No. 1, pp. 102–110

Singh, M., Kumar, M., Jain, R.K., & Chatrath, R.P. (1999). Radon in ground water related to seismic events. *Radiation Measurements,* Vol. 30, No. 4, pp. 465–469

Steinitz, G., Begin, Z.B., & Gazit-Yaari, N. (2003). Statistically significant relation between radon flux and weak earthquakes in the Dead Sea rift valley. *Geology,* Vol. 31, No. 6, pp. 505–508

Stranden, E., Kolstad, A.K., & Lind, B. (1984). Radon exhalation - moisture and temperature dependence. *Health Physics,* Vol. 47, No. 3, pp. 480–484

Thomas, D. (1988). Geochemical precursors to seismic activity. *Pure and Applied Geophysics,* Vol. 126, No. 2, pp. 241–266

Torkar, D., Zmazek, B., Vaupotič, J., & Kobal, I. (2010). Application of artificial neural networks in simulating radon levels in soil gas. *Chemical Geology,* Vol. 270, No. 1–4, pp. 1–8

Toutain, J.P., & Baubron, J.C. (1999). Gas geochemistry and seismotectonics: a review. *Tectonophysics,* Vol. 304, No. 1–2, pp. 1–27

Ulomov, V.I., & Mavashev, B.Z. (1967). On forerunners of strong tectonic earthquakes. *Doklady Akademii Nauk SSSR,* No. 176, pp. 319–322

Vàrhegyi, A., Hakl, J., Monnin, M., Morin, J.P., & Seidel, J.L. (1992). Experimental study of radon transport in water as test for a transportation microbubble model. *Journal of Applied Geophysics,* Vol. 29, No. 1, pp. 37–46

Virk, H.S., Sharma, A.K., & Sharma, N. (2002). Radon and helium monitoring in some thermal springs of North India and Bhutan. *Current Science,* Vol. 82, No. 12, pp. 1423–1424

Wakita, H., Nakamura, Y., Notsu, K., Noguchi, M., & Asada, T. (1980). Radon anomaly - possible precursor of the 1978 Izu-Oshima-Kinkai earthquake. *Science,* Vol. 207, No. 4433, pp. 882–883

Walia, V., Lin, S.J., Hong, W.L., Fu, C.C., Yang, T.F., Wen, K.L., & Chen, C.H. (2009a). Continuous temporal soil-gas composition variations for earthquake precursory studies along Hsincheng and Hsinhua faults in Taiwan. *Radiation Measurements,* Vol. 44, No. 9–10, pp. 934–939

Walia, V., Yang, T.F., Hong, W.L., Lin, S.J., Fu, C.C., Wen, K.L., & Chen, C.H. (2009b). Geochemical variation of soil-gas composition for fault trace and earthquake precursory studies along the Hsincheng fault in NW Taiwan. *Applied Radiation and Isotopes,* Vol. 67, No. 10, pp. 1855–1863

Washington, J.W., & Rose, A.W. (1990). Regional and temporal relations of radon in soil gas to soil-temperature and moisture. *Geophysical Research Letters,* Vol. 17, No. 6, pp. 829–832

Witten, I.H., & Frank, E. (1999). *Data Mining: Practical Machine Learning Tools and Techniques with Java Implementations.,* Morgan Kaufmann, San Francisco

Yang, T.F., Walia, V., Chyi, L.L., Fu, C.C., Chen, C.H., Liu, T.K., Song, S.R., Lee, C.Y., & Lee, M. (2005). Variations of soil radon and thoron concentrations in a fault zone and prospective earthquakes in SW Taiwan. *Radiation Measurements,* Vol. 40, No. 2–6, pp. 496–502

Zmazek, B., Italiano, F., Živčić, M., Vaupotič, J., Kobal, I., & Martinelli, G. (2002a). Geochemical monitoring of thermal waters in Slovenia: relationships to seismic activity. *Applied Radiation and Isotopes,* Vol. 57, No. 6, pp. 919–930

Zmazek, B., Todorovski, L., Džeroski, S., Vaupotič, J., & Kobal, I. (2003). Application of decision trees to the analysis of soil radon data for earthquake prediction. *Applied Radiation and Isotopes,* Vol. 58, No. 6, pp. 697–706

Zmazek, B., Todorovski, L., Živčić, M., Džeroski, S., Vaupotič, J., & Kobal, I. (2006). Radon in a thermal spring: Identification of anomalies related to seismic activity. *Applied Radiation and Isotopes,* Vol. 64, No. 6, pp. 725–734

Zmazek, B., Živčić, M., Todorovski, L., Džeroski, S., Vaupotič, J., & Kobal, I. (2005). Radon in soil gas: How to identify anomalies caused by earthquakes. *Applied Geochemistry,* Vol. 20, No. 6, pp. 1106–1119

Zmazek, B., Živčić, M., Vaupotič, J., Bidovec, M., Poljak, M., & Kobal, I. (2002b). Soil radon monitoring in the Krško Basin, Slovenia. *Applied Radiation and Isotopes,* Vol. 56, No. 4, pp. 649–657

10

Are There Pre-Seismic Electromagnetic Precursors? A Multidisciplinary Approach

Konstantinos Eftaxias

University of Athens, Faculty of Physics, Department of Solid State Section,
Panepistimiopolis Zofrafos, Athens
Greece

1. Introduction

In recent years, the wind prevailing in the scientific community does not appear to be favourable for earthquake (EQ) prediction research, in particular for the research of short term prediction [1]. Sometimes the arguments were extended to the extreme claim that any precursory activity is impossible [2]. Considering the difficulties associated with such factors as the highly complex nature, rarity of large EQs and subtleties of possible preseismic signatures, the present negative views are not groundless. It is difficult to prove associations between any two events (possible precursor and EQ) separated in time. To a certain extent, the aforementioned negative views were due to the fact that in the last decades the study of seismic precursors was expected to lead in a relatively short period of time to EQ prediction. However, the EQs are nothing but physical phenomena, and science should have some predictive power on their future behaviour of any physical system. In spite of this scepticism of the scientific community, the research towards the possible prediction of EQs in the future continues. This is attempted now with a more critical view taking into account new ideas and performing detailed theoretical, laboratory, field, and numerical investigations. Significant progress has been made in the research of precursory pattern changes of seismicity (e. g., Wyss and Martirosyan,[3]; Huang et al. [4]; Huang [5]) and the intermediate-term prediction of large EQs world-wide is already in the statistically proven stage (e g., Kossobokov et al. [6]). More recently, even the efforts to shorten the lead time to the "short-term" range are being made (e. g., Keilis-Borok et al.[7]). Some significant new waves have been rising in EQ science!

An EQ is a sudden mechanical failure in the Earth's crust, which has heterogeneous structures. The use of basic principles of fracture mechanics is a challenging field for understanding the EQ preparation process. A key fundamental question in strength considerations of materials is: *when does it fail?* Thus, a vital problem in material science and in geophysics is the identification of precursors of macroscopic defects or shocks. It is reasonable to expect that EQ's preparatory process has various facets which may be observed before the final catastrophe. *The science of EQ prediction should, from the start, be multidisciplinary!*

The present contribution focuses on fracture induced electromagnetic (EM) fields, which allow a real-time monitoring of damage evolution in materials during mechanical loading. Crack propagation is the basic mechanism of material failure. EM emissions in a wide frequency spectrum ranging from kHz to MHz are produced by opening cracks, which can be considered as the so-called precursors of general fracture. The radiated EM precursors are detectable both at a laboratory [8-16] and geological scale [17-37].

Data collection: Since 1994, a station has been installed and operated at a mountainous site of Zante island $(37.76^{0}N - 20.76^{0}E)$ in the Ionian Sea (western Greece). The main aim of this station is the detection of kHz-MHz EM precursors. Six loop antennas detect the three components (EW, NS, and vertical) of the variations of the magnetic field at 3 kHz and 10 kHz respectively; three vertical $\lambda/2$ electric dipoles detect the electric field variations at 41, 54 MHz, and 135 MHz respectively. These frequencies were selected in order to minimize the effects of the sources of man-made noise in the mountain area of the Zante Island. Moreover, two *Short Thin Wire Antennas*, oriented at EW and NS directions of length of 100 m, respectively, have been also installed. The aim of the last installation is the detection of ultra-low-frequency $(< 1Hz)$ EM precursors rooted in a preseismic lithosphere-atmosphere-ionosphere-coupling. All the EM time series were sampled at 1 Hz. Such an experimental setup helps to specify not only whether or not a single EM anomaly is preseismic in itself, but also whether a sequence of EM disturbances at different frequencies, which are emerged one after the other in a short time period, could be characterized as preseismic one. Clear such EM precursors have been detected over periods ranging from approximately a week to a few hours prior to catastrophic EQs that occurred in Greece or Italy (e.g., [21,22,25-37]). We emphasize that the detected precursors were associated with EQs: (i) occurred in land (or near coast-line); (ii) were strong, i.e., with magnitude 6 or larger; and (iii) were shallow. Recent results indicate that the recorded EM precursors contain information characteristic of an ensuing seismic event (e.g., [21,22,25-37]).

An important feature, observed both at laboratory and geophysical scale, is that it the MHz radiation precedes the kHz one [25,27-29,35,36]. Studies on the small (laboratory) scale reveal that the kHz EM emission is launched in the tail of pre-fracture EM emission from 97% up to 100% of the corresponding failure strength [25 and references therein]. At the geophysical scale the kHz EM precursors are emerged from a few days up to a few hours before the EQ occurrence. The association of MHz, kHz EM precursors with the last stages of EQ generation is justified.

The origin of EM emissions. The origin of EM emissions from fracture is not completely clear, and different attempts have been made in order to explain it [8, 32 and references therein]. A relevant attempt is related to the "capacitor model" [32]. In many materials, emission of photons, electrons, ions and neutral particles are observed during the formation of new surface. The rupture of inter-atomic (ionic) bonds also leads to intense charge separation, which is the origin of the electric charge between the micro-crack faces. On the faces of a newly created micro-crack the electric charges constitute an electric dipole or a more complicated system. The motion of a crack has been shown to be governed by a dynamical instability causing oscillations in its velocity and structure of the fractured surfaces. It is worth mentioning that laboratory experiments show that more intense fracto-emissions are observed during the unstable crack growth. Due to the crack strong wall vibration, in the stage of the micro-branching instability, it behaves as an efficient EM emitter [32].

Are there credible EM earthquake precursors? This is also a question debated in the science community. Despite fairly abundant circumstantial evidence, EM precursors have not been adequately accepted as real physical quantities [1]. There may be legitimate reasons for the critical views. The degree to which we can predict a phenomenon is often measured by how well we understand it. However, many questions about fracture processes remain standing. Especially, many aspects of EQ generation still escape our full understanding. Kossobokov [38] states that *"No scientific prediction is possible without exact definition of the anticipated phenomenon and the rules, which define clearly in advance of it whether the prediction is confirmed or not"*. We bear in mind that whether EM precursors to EQ exist is an important question not only for EQ prediction but also for understanding the physical processes of EQ generation. The comprehensive understanding of EM precursors in terms of physics is a path to achieve more sufficient knowledge of the last stages of the EQ preparation process and thus more sufficient short-term EQ prediction. A *seismic* shift in thinking towards basic science will lead to a renaissance of strict definitions and systematic experiments in the field of EQ prediction.

2. A proposed strategy for the study of MHz and kHz EM precursors

This chapter concentrates, in an appropriately critical spirit, on asking 3 crucial questions:

(i) How can we recognize an EM observation as a pre-seismic one?

(ii) How can we link an individual EM precursor with a distinctive stage of the earthquake preparation?

(iii) How can we identify precursory symptoms in EM observations which signify that the occurrence of the prepared EQ is unavoidable?

We shall attempt to approach the above mentioned questions in the simplest and most intuitive way, rather than emphasize mathematical rigor. In any case, the readers should be aware that this attempt refers to a *snap-shot* of a rapidly moving field.

One wonders whether necessary and sufficient criteria, have yet been established, that permit the characterization of an EM anomaly as a real EM precursor. One of the main purposes of this contribution is to suggest a procedure for the designation of observed kHz / MHz EM anomalies as seismogenic ones.

As it is said, an important feature, observed both at laboratory and geophysical scale, is that the MHz radiation precedes the kHz one [25, 28, 29 and references therein]. The remarkable asynchronous appearance of these precursors indicates that they refer to different stages of EQ preparation process. Moreover, it implies a different mechanism for their origin. Scientists ought to attempt to link the available various EM observations, which appear one after the other, to the consecutive processes occurring in Earth's crust.

The following *two stage model of EQ generation by means of pre-fracture EM activities* has been proposed: The pre-seismic MHz EM emission is thought to be due to the fracture of the highly heterogeneous system that surrounds the family of large high-strength entities distributed along the fault sustaining the system, while the kHz EM radiation is due to the fracture of the aforementioned large high-strength entities themselves [e.g.,28-30,32-36,39]. In the frame of the above mentioned two stage model, the identification of MHz and kHz EM precursors requires different methods of analysis.

2.1 Focus on MHz EM precursors

Fracture process in heterogeneous materials *can be attributed to phase transition of second order* [40,41,42]. This crucial property should be hidden in a seismogenic MHz EM activity [28,29,34,39]. The temporal evolution of a MHz EM precursor, which behaves as a second order phase transition, reveals transition from the phase from non-directional almost symmetrical cracking distribution to a directional localized cracking zone that includes the backbone of strong asperities (*symmetry breaking*) [29]. The identification of the time interval where the *symmetry breaking* is completed indicates that the fracture of heterogeneous system in the focal area has been obstructed along the backbone of asperities that sustain the system: *The siege of strong asperities begins [29]. However, the prepared EQ will occur if and when the local stress exceeds fracture stresses of asperities.* Consequently, the appearance of a really seismogenic MHz EM anomaly does not mean that the EQ is unavoidable (see Section 3).

2.2 Focus on kHz EM precursors

It has been suggested that the lounge of the kHz EM activity shows the fracture of asperities sustaining the fault [28,29,32-36]. This fracture is characterised by a non-equilibrium instability, thus acquiring a self-regulating character and to a great degree the property of irreversibility. The latter, is one of the most important components of prediction reliability. An associated fracto-EM precursor should show persistent behaviour and evolve as a phase transition far from equilibrium without any footprint of an equilibrium phase transition. Two questions effortlessly arise:

(i) *How can we recognize an observed kHz EM anomaly as a seismogenic one?*

(ii) *How does it indicate that the impending EQ is unavoidable?*

What follows concentrates on the above aforementioned two questions.

2.2.1 Statistical analysis of the kHz candidate EM precursors

An anomaly in a recorded time series is defined as a deviation from normal (background) behaviour. Concerning the development of a quantitative identification of kHz EM precursors, tools of information theory and concepts of entropy rooted in extensive and nonextensive statistical mechanics can be used in order to identify changes in the statistical pattern. A significant change is expected in the time series of the EM precursor, namely the appearance of entropy "drops" or information "peaks", revealing that the underlying fracto-EM mechanism is characterized by a high order of organization. The catastrophic fracture of asperities should be also characterized by a positive feedback mechanism. This means that the kHz EM precursors should show persistent behaviour (see Section 5.2.1).

2.2.2 Analysis in terms of universal structural patterns of fracture and faulting

From the early work of Mandelbrot [43], the aspect of self-affine nature of faulting and fracture is widely documented from field observations, laboratory experiments, and studies of failure precursors on the small (laboratory) and large (EQ) scale. The activation of a single fault should behave as a "reduced image" of the regional seismicity, and a "magnified image" of the laboratory seismicity. Moreover, fracture surfaces were found to be self-affine following the fractional Brownian motion (fBm) model over a wide range of length scales, while, the spatial roughness of fracture surfaces has been interpreted as a universal indicator of surface fracture, weakly dependent on the nature of the material and on the failure mode [27-30,35,36 and references therein]. Such universal structural patterns of fracture and faulting process should be included into an EM precursor which is rooted in the activation of a single fault. Therefore, an important pursuit is to examine whether universal patterns of fracture and faulting are hidden in the observed candidate kHz EM precursors (see Section 5).

2.2.3 Analysis by means of fractal electrodynamics

EQ's occur on a fractal structure of faults. An active crack or rupture, can be simulated by a "radiating element" [32]. The idea is that a fractal geo-antenna can be formed as an array of line elements having a fractal distribution on the focal area as the critical point is approached. The recently introduced Fractal Electrodynamics [44, 45], which combines fractal geometry with Maxwell's equations, offers a new possibility for the exploration of the kHz EM anomalies (see Section 7).

2.2.4 The science of EQ prediction should, from the start, be multi-disciplinary

EQ's preparatory process has various facets which may be observed before the final catastrophe. The science of EQ prediction should, from the start, be multidisciplinary. A candidate preseismic kHz-MHz EM activity should be consistent with other EM precursors (SES [46], EM precursors rooted in lithosphere-atmosphere-ionosphere coupling [47]) and precursors which are imposed by data from other disciplines such as: Seismology, Infrared Remote Sensing [48], Synthetic Aperture Radars Interferometry [49]. The sequential appearance of different precursors in a relative short time interval supports the seismogenic origin of each of them, increases the probability that a significant EQ is coming, and leads to higher estimation accuracy of its parameters, namely, magnitude, time and position (see Section 8). The EQ generation is a cooperative phenomenon and its prediction needs the cooperation of scientists!

2.2.5 Analysis in terms of complex systems

The field of study of complex systems holds that their dynamics is founded on universal principles that may be used to describe various crises [50,51]. The presence of common pathological symptoms in candidate kHz EM precursors on one hand and other catastrophic events (e.g., epileptic seizures, magnetic storms and solar flares), which clearly distinguish the catastrophic event from the corresponding normal state, strongly supports the seismogenic origin of the detected kHz EM anomalies (see Section 8.4).

The burden of this section was to describe a plausible scenario for the study of kHz EM precursors, without obvious internal inconsistencies and without violating the laws of physics. In the next sections we present results gained from previous studies applying the framework of analysis described above.

3. The precursory MHz EM activity as a second order phase transition phenomenon

In natural rocks at large length scales there are long-range anti-correlations, in the sense that a high value of a rock property, e.g. threshold for breaking, is followed by a low value and vice versa. Failure nucleation begins to occur at a region where the resistance to rupture growth has the minimum value. An EM event is emitted during this fracture. The fracture process continues in the same weak region until a much stronger region is encountered in its neighborhood. When this happens, fracture stops, and thus the emitted EM emission ceases. The stresses are redistributed, while the applied stress in the focal area increases. A new population of cracks nucleates in the weaker of the unbroken regions, and thus a new EM event appears, and so on. Therefore, the associated precursory MHz EM activity should be characterized by antipersistent behaviour and the interplay between the heterogeneities and the stress field should be responsible for this behaviour. This crucial feature is included in the recorded MHz EM precursors.

Physically, the presence of anti-persistency implies a set of EM fluctuations tending to induce stability to the system, essentially the existence of a non-linear negative feedback mechanism that "kicks" the opening rate of cracks away from extremes. The existence of such a mechanism leads to the next step: it has been proposed that the fracture of heterogeneous materials can be described in analogy with a continuous second order phase transition in equilibrium [40,41]. Thus, a seismogenic MHz EM activity, which is rooted in the fracture of the highly heterogeneous system that surrounds the family of large high-strength asperities, should be described as critical phenomenon. This critical signature is also hidden in the recorded MHz EM precursors [28-29,34-36,39]. The relevant analysis is based on the recently introduced Method of Critical Fluctuations (MCF) [52,53].

3.1 The method of critical fluctuations

The MCF, which constitutes a statistical method of analysis for the critical fluctuations in systems that undergo a continuous phase transition at equilibrium, has been recently introduced [52,53]. The authors have shown that the fluctuations of the order parameter ϕ, obey a dynamical law of intermittency which can be described in terms of a 1-d nonlinear map. The invariant density $\rho(\phi)$ for such a map is characterized by a plateau which decays in a super-exponential way (see Fig. 1 in [52]). For small values of ϕ, this critical map can be approximated as

$$\phi_{n+1} = \phi_n + u\phi_n^z + \epsilon_n \tag{1}$$

The shift parameter ϵ_n introduces a non-universal stochastic noise: each physical system has its characteristic "noise", which is expressed through the shift parameter ϵ_n. For thermal systems the exponent z is introduced, which is related to the isothermal critical exponent δ by $z = \delta + 1$.

The plateau region of the invariant density $\rho(\phi)$ corresponds to the laminar region of the critical map where fully correlated dynamics take place [29 and references therein]. The laminar region ends when the second term in Eq. (1) becomes relevant. However, due to the fact that the dynamical law (1) changes continuously with ϕ, the end of the laminar region cannot be easily defined based on a strictly quantitative criterion. Thus, the end of the laminar region should be generally treated as a variable parameter.

Based on the foregoing description of the critical fluctuations, the MCF develops an algorithm permitting the extraction of the critical fluctuations, if any, in a recorded time series. The important observation in this approach is the fact that the distribution $P(l)$ of the laminar lengths l of the intermittent map (1) in the limit $\epsilon_n \to 0$ is given by the power law [53]

$$P(l) \sim l^{-p_l} \tag{2}$$

where the exponent p_l is connected with the exponent z via $p_l = \frac{z}{z-1}$. Therefore the exponent p_l is related to the isothermal exponent δ by

$$p_l = 1 + \frac{1}{\delta} \tag{3}$$

with $\delta > 0$.

Inversely, the existence of a power law such as relation (2), accompanied by a plateau form of the corresponding density $\rho(\phi)$, is a signature of underlying correlated dynamics similar to critical behavior [52,53].

We emphasize that it is possible in the framework of universality, which is characteristic of critical phenomena, to give meaning to the exponent p_l beyond the thermal phase transitions [53].

The MCF is directly applied to time series or to segments of time series which appear to have a cumulative stationary behaviour. The main aim of the MCF is to estimate the exponent p_l. The distribution of the laminar lengths, l, of fluctuations included in a stationary window is fitted by the relation:

$$P(l) \sim l^{-p_2} e^{-p_3 l} \tag{4}$$

If p_3 is zero, then p_2 is equal to p_l. Practically, as p_3 approaches zero, then p_2 approaches p_l and the laminar lengths tend to follow a power-law type distribution. So, we expect a good fit to Eq. (4) with $p_2 > 1$ and $p_3 \approx 0$ if the system is in a critical state [50]. In terms of physics this behaviour means that the system is characterized by a "strong criticality", e.g., the laminar lengths tend to follow a power-law type distribution: during this critical time window the opening cracks (EM-emitters) are well correlated even at large distances [50].

We stress that when the exponent p_2 is smaller than one, then, independently of the p_3-value, the system is not in a critical state. Generally, the exponents p_2, p_3 have a competitive character, namely, when the exponent p_2 decreases the associated exponent p_3 increases (they are mirror images of each other). To be more precise, as the exponent p_2 ($p_2 < 1$) is close to 1 and simultaneously the exponent p_3 is close to zero, then the system is in a sub-critical state. As the system moves away from the critical state, then the exponent p_2 further decreases while simultaneously p_3 increases, reinforcing in this way the exponential character of the laminar length distribution: the EM fluctuations show short range correlations. In this way, we can identify the deviation from the critical state [50,52,53].

3.2 Application of the MCF method

On 13 May 1995 (8:47:13 UT) the Kozani-Grevena EQ (40.17°N, 21.68°E) occurred with magnitude $M = 6.6$. Fig. 1 shows the associated 41 MHz EM time series [25,28,29]. The data are sampled at 1 Hz.

A critical window (CW) has been identified including 23000 points (Fig. 1a) starting almost 11 hours before the time of the EQ occurrence. The corresponding distribution of the amplitude $P(\psi)$ of the emerged EM pulses in this CW is shown in Fig. (1c). It is characteristic the appearance of the plateau region in the top of distribution, as it is provided for the invariant density of critical map [52]. The laminar lengths l follow a power-law distribution $P(l)$. This feature indicated that the underlying fracture mechanism is characterized by fluctuations which are extended at many different time scales as well as the presence of long-range correlations. We note that the amplitude ψ_i of the preseismic MHz EM time series behaves as a kind of the order parameter [29]. Therefore, in the CW the fluctuations of the amplitude ψ_i of the recorded EM time series have an intermittent behaviour similar to the dynamics of the order parameter's fluctuations of a thermal critical system at the critical point. It is for this reason that this window is characterized as *critical window*.

A thermal phase transition is associated with a *symmetry breaking*. To gain inside into the temporal evolution of fracture, as the EQ is approaching, we elucidate the evolution of the *symmetry breaking* with time by making an analogy to a thermal continuous phase transition [29]. In the latter, the distribution of the fluctuations of the order parameter with temperature reveals the progress of the *symmetry breaking*. This distribution is almost a δ function at high temperature and evolves to a Gaussian with mean value zero as the system approaches the critical point. At the critical point, a characteristic plateau in the distribution appears, and the *symmetry breaking* emerges as the temperature further decreases. Below the critical temperature the distribution becomes again Gaussian, but its mean shifts to higher values associated with the *symmetry breaking*. As temperature approaches to $0°K$, where the *symmetry breaking* is completed, it becomes a δ function again. We look for these characteristic features in the preseismic time series, with stress taking on the role of temperature [29].

Let us look specifically at the precursor under study. Figs. (1b-1e) exhibit the distribution of the recorded EM fluctuations in successive time windows. As is was mentioned, the distribution of the amplitude (order parameter) in Fig. (1c) indicates the appearance of the CW. Fig. (1b) shows the distribution before the emergence of CW: the laminar lengths, l, do not follow a power-law-type distribution $P(l)$. The system is characterized by a spare almost symmetrical distributed in space random cracking with short-range correlations.

During the CW the sort-range correlation evolve to long-range; the corresponding distribution (Fig. 1c) might be considered as a precursor of the impending *symmetry breaking*. The *symmetry breaking* is readily observable in the subsequent time interval (Fig. 1d). The cracking is restricted in the narrow zone that includes the backbone of strong asperities distributed along the activated fault sustaining the system [29]. The distribution of the order parameter in Fig. (1e) is very similar to that of Fig. (1b). However, here there is an upward shift of the values to the range of the second lobe of the distribution in Fig. (1d). The laminar lengths does not follow a power-law distribution $P(l)$. The appearance of this window indicates that the *symmetry breaking* in the underlying fracto-EM process has been almost completed [29]. *The siege of strong asperities begins* [29]. However, the prepared EQ will occur if and when the local stress exceeds fracture stresses of asperities. The lounge of the kHz EM activity shows the fracture of asperities sustaining the fault [28,29,32-36]. Indeed, a very strong kHz EM burst appeared a few hours later and after that face the EQ occurred [29].

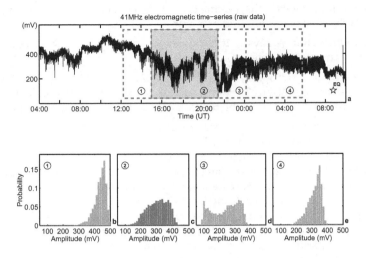

Fig. 1. The upper part shows the 41 MHz EM time series associated with the Kozani-Grevena EQ. The lower part elucidates the evolution of symmetry breaking with time.

4. How can we recognize a kHz EM anomaly as a pre-seismic one?

An anomaly in a recorded time series is defined as a deviation from normal (background) behaviour. In order to develop a quantitative identification of EM precursors, tools of information theory and concepts of entropy are used in order to identify statistical patterns. Entropy and information are seen to be complementary quantities, in a sense: entropy "drops" have as a counterpart information "peaks" in a more ordered state. The seismicity is a critical phenomenon [41,54] , thus, it is expected that a significant change in the statistical pattern, namely the appearance of entropy "drops" or information "peaks", represents a deviation from normal behaviour, revealing the presence of an EM anomaly.

It is important to note that one cannot find an optimum organization or complexity measure. Thus, a combination of some such quantities which refer to different aspects, such as structural or dynamical properties, is the most promising way.

Several well-known techniques have been applied to extract EM precursors hidden in kHz EM time series:

(i) *T*-entropy: It is based on the intellectual economy one makes when rewriting a string according to some rule [55].

(ii) Approximate entropy: It provides a measure of the degree of irregularity or randomness within a series of data. More precisely, this examines the presence of similar epochs in time series; more similar and more frequent epochs lead to lower values of approximate entropy [35 and references therein].

(iii) Fisher Information: It represents the amount of information that can be extracted from a set of measurements [56].

(iv) Correlation Dimension: It measures the probability that two points chosen at random will be within a certain distance of each other, and examines how this probability changes as the distance is increased [57].

(v) R/S analysis: It provides a direct estimation of the Hurst Exponent which is a precious indicator of the state of randomness of a time-series [58].

(vi) Detrended Fluctuation Analysis: It has been proven useful in revealing the extent of long-range correlations in time series [59, 60].

(vii) Shannon n-block entropies (conditional entropy, entropy of the source, Kolmogorov-Sinai entropy): They measure the uncertainty of predicting a state in the future, provided a history of the present state and the previous states [61-65].

(viii) Tsallis entopy: One of the crucial properties of the Boltzmann-Gibbs entropy in the context of classical thermodynamics is extensivity, namely proportionality with the number of elements of the system. The Boltzmann-Gibbs (B-G) entropy satisfies this prescription if the subsystems are statistically (quasi-) independent, or typically if the correlations within the system are essentially local. In such cases the energy of the system is typically extensive and the entropy is additive. In general, however, the situation is not of this type and correlations may be far from negligible at all scales. Inspired by multifractals concepts, Tsallis [66, 67] has proposed a generalization of the B-G statistical mechanics. He introduced an entropic index q which leads to a nonextensive statistics. The value of q is a measure of the nonextensivity of the system: $q = 1$ corresponds to the standard, extensive, B-G statistics. The order of organization of the nonextensive systems is measured by the Tsallis entropy.

The application of all the above mentioned multidisciplinary statistical procedure [30,33,35,36,68-71] sensitively recognizes and discriminates the candidate EM precursors from the EM background: they are characterized by significantly higher organization in respect to that of the EM noise in the region of the station. However, we should keep in mind that though a sledge hammer may be wonderful for breaking rock, it is a poor choice for driving a tack into a picture frame!

5. Focus on the possible seismogenic origin of the detected kHz EM anomaly by means of universally holding scaling laws of fracture

As it is mentioned in the previous Section, all the applied techniques reveal that the kHz EM anomaly is characterized by a significant lower complexity (or higher organization). Importantly this anomaly is also characterized by strong persistency [28,29]. The simultaneous appearance of both these two crucial characteristics implies that the underlying fracture process is governed by a positive feedback mechanism which is consistent with an anomaly being a precursor of an ensuing catastrophic event.

However, we suggest that any multidisciplinary statistical analysis by itself is not sufficient to characterize an emerged kHz EM anomaly as a pre-earthquake one. Much remains to be done to tackle systematically real pre-seismic EM precursors.

As it is mentioned in Section 2.2, the Earth's crust is extremely complex. However, despite its complexity, there are several universally holding scaling relations. Such universal structural patterns of fracture and faulting process should be included into an EM precursor which is rooted in the activation of a single fault. Therefore an important pursuit is to investigate whether universal features of fractures and faulting are included in the recorded kHz EM precursors.

5.1 The activation of a single fault as a self-affine image of the regional and laboratory seismicity

The self-affine nature of faulting and fracture predicts that the activation of a single fault is a reduced / magnified image of the regional/ laboratory seismicity, correspondingly (see Section 2.2.2). A fracto-EM precursor rooted in the activation of a single fault should be consistent with the above mentioned requirement.

5.1.1 The activation of a single fault as a "reduced self-affine image" of the regional seismicity

A model for EQ dynamics coming from a non-extensive Tsallis formulation [66,67] has been recently introduced by Sotolongo-Costa and Posadas, [72]. Silva et al. [73] have revised this model. The authors assume that the mechanism of relative displacement of fault plates is the main cause of EQs. The space between fault planes is filled with the residues of the breakage of the tectonic plates, from where the faults have originated. The motion of the fault planes can be hindered not only by the overlapping of two irregularities of the profiles, but also by the eventual relative position of several fragments. Thus, the mechanism of triggering EQs is established through the combination of the irregularities of the fault planes on one hand and the fragments between them on the other hand. This nonextensive approach leads to a Gutenberg-Ricter (G-R) type law for the magnitude distribution of EQs:

$$\log(N_{>m}) = \log N + \left(\frac{2-q}{1-q}\right) \log\left[1 - \left(\frac{1-q}{2-q}\right)\left(\frac{10^{2m}}{\alpha^{2/3}}\right)\right] \tag{5}$$

where N is the total number of EQs, $N(> m)$ the number of EQs with magnitude larger than m. Parameter α is the constant of proportionality between the EQ energy, ε and the size of fragment. The entropic index q describes the deviation of Tsallis entropy from the traditional Shannon one. The proposed non-extensive G-R type law (5) provides an excellent fit to seismicities generated in various large geographic areas, each of them covering many geological faults. We emphasize that the q-values are restricted in the narrow region from 1.6 to 1.8 [72-74]. Notice, the magnitude-frequency relationship for EQs do not say anything about a specific activated fault (EQ). A kHz EM precursors refers to the activation of a specific fault. Thus, we examine whether the kHz EM activity also follows the Eq. (5).

Definition of the "Electromagnetic earthquake": We regard as amplitude A of a candidate "fracto-EM fluctuation" the difference $A_{fem}(t_i) = A(t_i) - A_{noise}$, where A_{noise} is the background (noise) level of the EM time series. We consider that a sequence of k successively emerged "fracto-EM fluctuations" $A_{fem}(t_i)$, $i = 1, \ldots, k$ represents the EM energy released, ε, during the damage of a fragment. We shall refer to this as an "electromagnetic earthquake" (EM-EQ). Since the squared amplitude of the fracto-EM emissions is proportional to their energy, the magnitude m of the candidate "EM-EQ" is given by the relation

$$m = \log \varepsilon \sim \log\left(\sum \left[A_{fem}(t_i)\right]^2\right) \tag{6}$$

The Eq. (5) provides an excellent fit to the pre-seismic kHz EM experimental data incorporating the characteristics of nonextensivity statistics into the distribution of the detected precursory "EM-EQs" [32,33,36,75]. Herein, $N(> m)$ is the number of "EM-EQs" with magnitude larger than m, $P(> m) = N(> m)/N$ is the relative cumulative number of "EM-EQs" with magnitude larger than m, and α is the constant of proportionality between the EM energy released and the size of fragment. The best-fit q parameter for this analysis has been estimated to be approximately 1.8 [32,33,36,75].

It is very interesting to observe the similarity in the q-values associated with the non-extensive Eq. (5) for: (i) seismicities generated in various large geographic areas, and (ii) the precursory sequence of "EM-EQs". This finding indicates that the statistics of regional seismicity could be merely a macroscopic reflection of the physical processes in the EQ source, as it has been suggested by Huang and Turcotte [76].

5.1.2 The activation of a single fault as a "magnified self-affine image" of the laboratory seismicity

Rabinovitch et al. [77] have studied the fractal nature of EM radiation induced in rock fracture. The analysis of the prefracture EM time series reveals that the cumulative distribution of the amplitudes also follows a power law with exponent $b = 0.62$. A similar statistical analysis of kHz EM precursor associated with the Athens EQ reveals that this also follows the power law $N(> A) \sim A^{-b}$, where $b = 0.62$ [78].

In seismology, a well known scaling relation between magnitude and the number of EQs is given by the Gutenberg-Richter (G-R) relationship:

$$\log N(> M) = \alpha - bm \tag{7}$$

where, $N(> M)$ is the cumulative number of EQs with a magnitude greater than M occurring in a specified area and time and b and α are constants.

Importantly, the Gutemberg-Ricther law also holds for acoustic emission events in rock samples [79]. Laboratory experiments by means of acoustic emissions also show a significant decrease in the level of the observed b-values immediately before the global fracture [79]. Characteristically, Ponomarev et al. [80] have reported a significant fall of the observed b-values from ~ 1 to ~ 0.6 just before the global rupture. Recently, Lei and Satoh [81], based on acoustic emission events recorded during the catastrophic fracture of typical rock samples under differential compression, suggest that the pre-failure damage evolution is characterized by a dramatic decrease in b-value from ~ 1.5 to ~ 0.5 for hard rocks. There are increasing reports on premonitory decrease of b-value before EQs: foreshock sequences and main shocks are characterized by a much smaller exponent compared to aftershocks [82]. We emphasise the sequence of kHz EM-EQs associated with the Athens EQ also follows the Gutenberg-Richter law with $b = 0.51$ [32].

The above mentioned results verify that the activation of a single fault behaves as a magnified self-affine image of the laboratory seismicity and reduced image of the regional seismicity.

5.2 Signatures of fractional-Brownian-motion nature of faulting and fracture in the candidate kHz EM precursor

Fracture surfaces were found to be self-affine following the fractional Brownian motion model (see Section 2.2.2) [27-30,35,36 and references therein]. This universal feature should be included into an kHz EM precursor. If a time series is a temporal fractal then a power-law of the form $S(f) \propto f^{-\beta}$ is obeyed, with $S(f)$ the power spectral density and f the frequency. The spectral scaling exponent β is a measure of the strength of time correlations. The goodness of the power-law fit to a time series is represented by a linear correlation coefficient, r. Based on a fractal spectral analysis, which has been performed by means of wavelets, it has been shown [27-30,35,36] that the emergent strong kHz EM precursors follow the law $S(f) \propto f^{-\beta}$; the coefficient r takes values very close to 1, i.e., the fit to the power-law is excellent. This result shows the fractal character of the underlying processes and structures.

The β exponent takes high values, i.e., between 2 and 3. This finding reveals that:

(i) The EM bursts have long-range temporal correlations, i.e. strong memory: the current value of the precursory signal is correlated not only with its most recent values but also with its long-term history in a scale-invariant, fractal manner.

(ii) The spectrum manifests more power at lower frequencies than at high frequencies. The enhancement of lower frequency power physically reveals a predominance of larger fracture events. This footprint is also in harmony with the final step of EQ preparation.

(iii) Two classes of signal have been widely used to model stochastic fractal time series, fractional Gaussian (fGn) and fractional Brownian motion (fBm) model [83]. The fGn-model the scaling exponent β lies between -1 and 1, while the fBm regime is indicated by β values from 1 to 3. The estimated β exponent successfully distinguishes the candidate precursory activities from the EM noise [27-31,35,36]. Indeed, the β values in the EM background are between 1 and 2 indicating that the time profile of the EM series during the quiet periods is qualitatively analogous to the fGn class. On the contrary, the β values in the candidate EM precursors are between 2 and 3, suggesting that they belong to the fBm class.

In summary, the fBm nature of faulting and fracture is included in the kHz EM precursors.

5.2.1 Persistent behaviour of the detected kHz EM precursors

The β exponent is related to the Hurst exponent H by the formula [83, 84]:

$$\beta = 2H + 1 \tag{8}$$

with $0 < H < 1$ $(1 < \beta < 3)$ for the fractional Brownian motion (fBm) model. The exponent H characterizes the persistent / anti-persistent properties of the signal. The range $0.5 < H < 1$ $(2 < \beta < 3)$ indicates persistency, which means that if the amplitude of the fluctuations increases in a time interval it is likely to continue increasing in the next interval. We recall that we found β values in the candidate EM precursors to lie between 2 and 3. The H values are close to 0.7 in the strong segments of the kHz EM activity [27-31,35,36]. This means that the EM fluctuations are positively correlated: the underlying dynamics is governed by a positive feedback mechanism. External influences would then tend to lead the system out of equilibrium. The system acquires a self-regulating character and, to a great extent, the property of irreversibility, one of the important components of prediction reliability. Sammis and Sornette [85] have recently presented the most important positive feedback mechanisms.

It is expected that a positive feedback mechanism results in a finite-time singularity. The kHz EM time series under study shows such a behaviour by means of the "cumulative Benioff type EM energy release". A clear finite-time singularity of this type has been reported in [27,28,78].

Remark: The estimated Hurst exponents through the R/S analysis are in harmony with those estimated from the fractal spectral analysis via the hypothesis that the time series follows the fBm-model [35,36]. This fact supports the hypothesis that the profile of kHz EM precursors follow the persistent fBm-model. The last hypothesis has been further verified by a DFA-analysis [35,36].

5.3 Footprints of universal roughness value of fracture surfaces in the kHz EM activity

The Hurst exponent, H, specifies the strength of the irregularity ("roughness") of the fBm surface topography: the fractal dimension is calculated from the relation $D = (2 - H)$ [83].

The Hurst exponent $H \sim 0.7$ has been interpreted as a universal indicator of surface fracture, weakly dependent on the nature of the material and on the failure mode [86-90]. Importantly, the surface roughness of a recently exhumed strike-slip fault plane has been measured by three independent 3D portable laser scanners [91]. Statistical scaling analyses show that the striated fault surface exhibits self-affine scaling invariance that can be described by a scaling roughness exponent, $H_1 = 0.7$ in the direction of slip. In Section 5.2.1 we showed that the "roughness" of the profile of the kHz EM precursors, as it is represented by the Hurst exponent, is distributed around the value 0.7. This result has been verified by means of both fractal spectral analysis and R/S analysis [35,36]. Thus, the universal spatial roughness of fracture surfaces nicely coincides with the roughness of the temporal profile of the recorded kHz EM precursors.

6. Interpretation of precursory kHz EM activity in terms of Intermittent Criticality

The Intermittent Criticality (IC)-viewpoint of EQ dynamics is based on the hypothesis that a large regional EQ is the end result of a process in which the stress field becomes correlated over increasingly long scale-lengths, which set the size of the largest EQ that can be expected at any given time. The largest event on the fault network cannot occur until regional criticality has been achieved and stress is consequently correlated at all length scales up to the size of the region. The growth of the spatial correlation length obeys a power law with a singularity in the critical point [92-102]. This large event destroys, after its occurrence, the criticality on its associated network, creating a period of relative quiescence, after which the process repeats by rebuilding correlation lengths towards criticality and the next large event. In contrast to self-organized criticality, in which the system is always at or near criticality, intermittent criticality implies time-dependent variations in the activity during a seismic cycle. Before the large EQ, the growing correlation length manifests itself as an increase in the frequency of intermediate-magnitude earthquakes. This is commonly referred to as the "accelerating moment release model", and has been discussed by a number of authors [97,98]. Briefly, IC-approach includes self-organized criticality, growing spatial correlation length, and accelerating energy release.

A kHz EM anomaly can be interpreted as an EM confirmation of the IC-hypothesis. Indeed, a power-law type increase in the rate of EM energy release as the global instability approaches is observed [27,28,78]. The recorded acceleration of the EM emission leading up to EM large event and "EM shadow" following this is in harmony with the IC-hypothesis. Notice, the rate of seismic energy release computed around the epicenter of the EQ follows a similar power-law type increase [27,28,78]. This experimental fact supports the hypothesis that both the seismicity and the preseismic EM activity represent two cuts in the same underlying fracture mechanism. Moreover, the spectral scaling exponent β (see Section 5.2) is a measure of the strength of time correlations. The β-values are significantly shifted to higher values as the EQ is approaching [27,28,78], namely, the correlation length in the time series increases as the catastrophic event approaches. Consequently, the two basic signatures predicted by the IC-model are included in the candidate kHz EM precursors.

7. Interpretation of MHz-kHz EM precursors in terms of fractal electrodynamics

Recently, the research area known as "fractal electrodynamics" has been established. This term was first suggested by Jaggard [44,45] to identify the newly emerging branch of research, which combines fractal geometry with Maxwel's theory of electrodynamics. From the laboratory scale to the geophysical scale, fault displacements, fault and fracture trace length, and fracture apertures follow a power-law distribution. Thus a fault shows a fractal pattern: a network of line elements having a fractal distribution in space is formed as the event approaches. However, an active crack or rupture can be simulated as a radiating element. The idea is that a *Fractal EM Geo-Antenna* can be formed as an array of line elements having a fractal distribution on the ground surface as the significant EQ is approached. This idea has been tested in [27]: the precursors are governed by characteristics (e.g., scaling laws, temporal evolution of the spectrum content, broad-band spectrum region, and accelerating emission rate) predicted by fractal electrodynamics. Notice, the fractal tortuous structure can significantly increase the radiated power density, as compared to a single dipole antenna. The tortuous path increases the effective dipole moment, since the path length along the emission is now longer than the Euclidean distance, and thus the possibility to capture these preseismic radiations by aerial antennas.

The fractal dimension of the *fractal EM geo-antenna* associated with the Athens EQ is $D = 1.2$ [27]. Seismological measurements as well as theoretical studies [101,102 and references therein] suggest that a surface trace of a single major fault might be characterized by $D = 1.2$. We clarify that the exponent D does not describe the geometrical setting of the rupture faults but it only gives the distribution of rupture fault lengths irrespective of their positions. More information is needed for a full geometrical interpretation of the faults, e.g. the position of the rupture centers.

8. The science of EQ prediction should, from the start, be multi-disciplinary

As it was mentioned in Introduction, EQ's preparatory process has various facets which may be observed before the final catastrophe, thus, a candidate preseismic EM activity should be consistent with other EM precursors or precursors that are imposed by data from other disciplines (Seismology, Infrared Remote Sensing, Synthetic Aperture Radars Interferometry e.t.c.).

8.1 Seismic electric signals

A well documented type of precursory signals is the so-called seismic electric signals (SES) [103]. They are transient low frequency ($< 1Hz$) electric signals and are consistent with the "pressure stimulated currents model", which suggests that, upon a gradual variation of the pressure (stress) on a solid, transient electric signals are emitted, from the reorientation of electric dipoles formed due to disorder in the focal area, when approaching a critical pressure. Field and laboratory experience coincide to the point that the transient SES tend to appear earlier in respect to the MHz-kHz EM precursors. In a recent paper, Varotsos et al. [54] report that the occurrence time of a main shock is specified in advance by analyzing in "natural time" the seismicity subsequent to the initiation of the SES activity. This analysis identifies the time when the seismicity approaches the critical state. The authors conclude that, from the time of that critical state, "the main shock was found empirically to follow usually within a few days up to one week". It is important to note that: (i) MHz / kHz EM precursors are emerged from approximately a week up to a few hours before the EQ occurence, namely, when the earth crust is in critical state by means of seismicity. (ii) MHz EM precursors can also be attributed to a phase transition of second order, as it happens for the seismicity preceding main shocks. Bear in mind that, in the frame of the proposed two stage model, MHz EM precursors are rooted in fracture of heterogeneous regime which surrounds the activated fault. The finally emerged kHz EM precursors indicate that the occurrence of the prepared EQ is unavoidable. This scheme, namely, the appearance of SES following by kHz-MHz EM precursory radiations, has been reported before EQs that occurred in Greece [21,25,30,104]. We note that, using Fisher Information and entropy metrics, it has been found that both the organization of the seismicity around the activated fault and the organization of the kHz EM precursors significantly increase as the EQ approaches [105].

8.2 EM anomalies rooted in preseismic LAI-coupling

A class of precursors is rooted in anomalous propagation of EM signals over epicentral regions due to a pre-seismic Lithosphere-Atmosphere-Ionosphere (LAI)coupling [1 and references therein]. During quiet periods, the daily EM data present a main bay-like behaviour. The records refer to the Earth-ionosphere waveguide propagation of natural EM emissions. Any change in the lower ionosphere due to an induced pre-seismic LAI-coupling may result in significant changes in the signal propagation-received at a station. Therefore, the emergence of an ionospheric EM anomaly is recognized by a strong perturbation of the characteristic bay-like morphology in the chain of daily data. Pulinets et al. [106] have reported that ionospheric precursors within 5 days before the seismic shock are registered in 100% of the cases for EQs with magnitude 6 or larger. Such anomalies have been recorded in Greece [21, 27, 104]. *Importantly, these anomalies were followed by well documented preseismic sequence of MHz and kHz EM activities, while SES appeared earlier.* The EM precursors sourced in the preseismic LAI-coupling and the MHz/kHz EM precursors appear during the last days before the main shock, namely, when the earth crust was in critical state by means of seismicity.

8.3 Precursors imposed by data from other disciplines

As it was emphasized in Introduction, EQ's preparatory process has various facets which may be observed before the final catastrophe. On September 7, 1999 the catastrophic Athens (Greece) EQ with a magnitude $M_w = 5.9$ occurred. The following sequence of well documented different precursors have been observed [26,29,30,104]:

1. A clear SES activity was recorded.

2. MHz EM anomalies were simultaneously recorded at 41, 54, and 135 MHz on August 29, 1999. These anomalies can be attributed to a phase transition of second order by means of the analysis reported in Section 3.

3. Two stong burst-like EM anomalies at 3 and 10 kHz were simultaneously recorded before the EQ occurrence. The first and second anomaly lasted for 12 and 17 hours, respectively, with a cessation of 9 hours. The second anomaly ceased about 9 hours before the EQ. This preseismic activity obeys all the requirements of the Section 2.2.

4. Infrared remote sensing makes use of satellite infrared sensors to detect infrared radiation emitted from the Earth's surface before EQs. A clear increase in the thermal infrared radiation (TIR) over the area around the Athen's EQ epicentre recorded during the last days before the EQ. The appearance of TIR emissions enhances the consideration that the fracture process has been extended up to the surface layers of the crust in the case of this EQ.

5. Synthetic aperture radars (SAR) are space-borne instruments that emit EM radiation and then record the strength and time delay of the returning signal to produce images of the ground. By combining two or more SAR images of the same area, it is possible to generate elevation maps and surface change maps with unprecedented precision and resolution. This technique is called SAR interferometry. SAR interferometry is becoming a new tool for active tectonics by providing both mm-precision surface change maps spanning periods of days to years and m-precision, high-resolution topographic maps for measuring crustal strain accumulated over longer periods of time. The fault modelling of the Athens EQ, based on information obtained by radar interferometry (ERS-2 satellite), predicts two faults: the main fault segment is responsible for 80% of the total energy released, with the secondary fault segment for the remaining 20%. A recent seismic data analysis carried out by Kikuchi, using the now standard methodology, also indicates that a two-event solution for the Athens EQ is more likely than a single event solution. According to Kikuchi, there was probably a subsequent ($M = 5.5$) EQ after about 3.5 s of the main event ($M = 5.8$). On the other hand, two strong impulsive kHz EM bursts were emerged in the tail of the preseismic EM emission. The first burst contains approximately 20% of the total EM energy received during the emergence of the two bursts, and the second the remaining 80%. The aforementioned surprising correlation in the energy domain between the two strong preseismic kHz EM signals and two faults activated, strongly supports beyond any analysis the hypothesis that the two strong EM bursts reveal the nucleation of the impending EQ.

6. A precursory power-law-type acceleration of the seismic energy release has been observed in the case of Athens EQ. The apparent onset of precipitous power-law behaviour began approximately 20 days before the EQ and culminated with the main event appearance, disappearing soon afterward.

The aforementioned observed phenomena, support the proposal that *"the science of EQ prediction should, from the start, be multi-disciplinary!"*

8.4 Universality among various geophysical and biological catastrophic events

In the last 20 years, the study of complex systems has emerged as a recognized field in its own right, although, a good definition of what a complex system is has proven elusive. Is there a common factor in the seemingly diverse complex phenomena? The answer is yes-they happens in systems consisting of many similar units interacting in a relatively well-defined manner; the field of study of complex systems holds that their dynamics is founded on universal principles that may be used to describe phenomena that are otherwise quite different in nature. When one considers a phenomenon or a thing that is "complex", one generally associates it with something that is *hard to separate, analyze or to solve*. Instead, we refer to a complex system as one whose phenomenological laws, which describe the global behaviour of the system, are not necessarily directly related to the microscopic laws that regulate the evolution of its elementary parts. The main features of this collective bahaviour are that an individual unit's action is dominated by the influence of its neighbours, the unit behaves differently from the way it would behave on its own; and such systems show ordering phenomena as the units simultaneously change their behaviour to a common pattern [107-109]. Generally, topological disorder within the complex system introduces new, surprising effects, the laws that describe their behaviour are qualitatively different from those that governs its units. Therefore, the description of the entire system's behaviour requires a qualitatively new theory. Interesting principles have been proposed in an attempt to provide such a unified theory. These include self-organization, intermittent criticality, simultaneous existence of many degrees of freedom, self-adaption, rugged energy landscapes, and scaling (for example, power-law dependence) of the parameters and the underlying network of connections.

Empirical evidence has been mounting that supports the possibility that a number of systems arising in disciplines as diverse as physics, biology, engineering, and economics may have certain quantitative features that are intriguingly similar. Picoli et. al. [110] reported similarities between the dynamics of geomagnetic signals and heartbeat intervals. de Arcangelis et al. [111] presented evidence for universality in solar flare and earthquake occurrence. Kossobokov and Keilis-Borok [112] have explored similarities of multiple fracturing on a neutron star and on the Earth, including power-law energy distributions, clustering, and the symptoms of transition to a major rupture. Sornette and Helmstetter [113] have presented occurrence of finite-time singularities in epidemic models of rupture, EQs, and starquakes. Abe and Suzuki [114] have shown that internet shares with EQs common scale-invariant features in its temporal behaviours. Peters et al. [115] have shown that the rain events are analogous to a variety of nonequilibrium relaxation processes in nature such as EQs and avalanches. Fukuda et al. [116] have shown similarities between communication dynamics in the Internet and the automatic nervous system.

A corollary in the study in terms of complexity is that transferring ideas and results from investigators in hitherto disparate areas will cross-fertilize and lead to important new results. Considering the rarity of large surface EQs which occurs on land and subtleties of possible preseismic EM signatures, the study of EM precursors by means of complexity offers new possibilities for their exploration.

Importantly, the strong analogies between the dynamics of EQ and neurobiology have been realized by numerous authors [117-123]. In general, authors have suggested that EQ's dynamics and neurodynamics can be analyzed within similar mathematical frameworks [117-127]. Characteristically, driven systems of interconnected blocks with stick-slip friction capture the main features of EQ process. These models, in addition to simulating the aspects of EQs may also represent the dynamics of neurological networks [117 and references therein]. Hopfield [118] proposed a model for a network of N integrate-and-fire neurons. In this model, the dynamical equation of kth neuron equation 28 in [118] is based on the Hodgekin-Huxley model for neurodynamics and represents the same kind of mean field limit that has been examined in [123], in connection with EQs.

Recently, it has been shown that a unified approach to catastrophic events-from the normal state of earth / brain to EQ by means of preseismic kHz EM emission/epileptic seizure exists.

The appearance of common "pathological" symptoms, i.e, high organization, persistency, and acceleratin energy release accompanies the emergence of kHz EM precursors and seizures [124-126]. More recently, Osorio et al. [127] have shown that a dynamical analogy supported by five scale-free statistics , namely, the Gutenberg-Richter distribution of event sizes, the distribution of interevent intervals, the Omori and inverse Omori laws, and the conditional waiting time until the next event, is shown to exist between seizures and EQs.

Strong analogies between the dynamics of kHz EM precursors and that and magnetic storms have been realized. The appearance of common "pathological" symptoms, i.e, high organization, persistency, and accelerating energy release accompanies the emergence of these two crises [128-131]. Moreover, the Tsallis-based energy distribution function (Eq. 5) is able to describe solar events and magnetic storms, as well. The best-fit for this analysis is given by a q-parameter value equal 1.82 and 1.84, correspondingly [131]. It is very interesting to observe the similarity in the q-values for: (i) seismicities generated in various large geographic areas, (ii) the precursory sequence of "EM-EQs" associated with the activation of a single fault,(iii) solar flares, and (iv) magnetic storms. This experimental evidence could be considered as an indication of universality among various geophysical processes. A unified theory may exist for the ways in which the above mentioned different systems organize themselves to produce a large geological or biological crisis.

9. Conclusions

As mentioned in Introduction, a key question debated in the scientific community is: Are there credible EM earthquake precursors? Despite fairly abundant circumstantial evidence, EM precursors have not been adequately accepted as real physical quantities, and there may be legitimate reasons for the critical views. In this contribution we propose a strategy for the study of MHz and kHz EM precursors which concentrates in an appropriately critical spirit, on asking 3 crucial questions:

(i) How can we recognize an EM observation as a pre-seismic one?

An anomaly in a recorded time series is defined as a deviation from normal (background) behaviour. In order to develop a quantitative identification of EM precursors, tools of information theory and concepts of entropy are used in order to identify statistical patterns. Entropy and information are seen to be complementary quantities, in a sense: entropy "drops" have as a counterpart information "peaks" in a more ordered state. The seismicity is a critical phenomenon [41,54], thus, it is expected that a significant change in the statistical pattern, namely the appearance of entropy "drops" or information "peaks", represents a deviation from normal behaviour, revealing the presence of an EM anomaly. Several well-known techniques have been applied to extract EM precursors hidden in kHz EM time series: T-entropy, Approximate entropy, Fisher Information, Correlation Dimension, R/S analysis, Detrended Fluctuation Analysis, Shannon n-block entropies (conditional entropy, entropy of the source, Kolmogorov-Sinai entropy), Tsallis entopy. It is important to note that one cannot find an optimum organization or complexity measure. Thus, a combination of some such quantities which refer to different aspects, such as structural or dynamical properties, is the most promising way. The application of all the above mentioned multidisciplinary statistical procedure [30,33,35,36,69-71] sensitively recognizes and discriminates the candidate kHz EM precursors from the EM background: they are characterized by significantly higher organization / lower complexity in respect to that of the EM noise in the region of the station. Importantly this pre-seismic EM emission is also characterized by strong persistency [28,29]. The simultaneous appearance of both these two crucial characteristics, i.e., higher organization and persistency, implies that the underlying fracture process is governed by a positive feedback mechanism which is consistent with an anomaly being a precursor of an ensuing catastrophic event.

However, we suggest that any multidisciplinary statistical analysis by itself is not sufficient to characterize an emerged kHz EM anomaly as a pre-earthquake one. Much remains to be done to recognise systematically real pre-seismic EM precursors. The Earth's crust is clearly extremely complex. However, despite its complexity, there are several universally valid scaling relations. From the early work of Mandelbrot, much effort has been put to statistically characterise the resulting fractal surfaces in fracture processes. Fracture surfaces were found to be self-affine following the fractional Brownian motion (fBm) model over a wide range of length scales. Moreover, the spatial roughness of fracture surfaces has been interpreted as a universal indicator of surface fracture, weakly dependent on the nature of the material and on the failure mode. The Hurst Exponent H specifies the strength of the irregularity ("roughness") of the surface topography and the value of $H \sim 0.7$ has been interpreted as a universal indicator of surface fracture, weakly dependent on the nature of the material and the failure mode. Therefore, an important pursuit is to make a quantitative comparison between fractal patterns possibly hidden in an emergent kHz EM anomaly on one hand and universal fractal patterns of fracture surfaces on the other hand: an EM precursor associated with the last stage of EQ generation should behave as a persistent fBm temporal fractal, while the "roughness" of its profile, as it is represented by the Hurst exponent, should be characterized by the value $H \sim 0.7$. These two universal features of fracture are hidden in the recorded kHz EM precursors (see Section 5).

The self-affine nature of faulting and fracture predicts that the activation of a single fault is a reduced / magnified image of the regional/laboratory seismicity, correspondingly [76]. A fracto-EM precursor rooted in the activation of a single fault should be consistent with the above mentioned requirement. The sequence of kHz "electromagnetic earthquakes" rooted in the activation of a single fault satisfies the aforementioned requirement.

(ii) How can we link an individual EM precursor with a distinctive stage of the earthquake preparation?

An important feature, observed both at laboratory and geophysical scale, is that the MHz radiation precedes the kHz one. The remarkable asynchronous appearance of these precursors indicates that they refer to different stages of EQ preparation process. Moreover, it implies a different mechanism for their origin. Scientists ought to attempt to link the available various EM observations, which appear one after the other, to the consecutive processes occurring in Earth's crust.

The following *two stage model of EQ generation by means of pre-fracture EM activities* has been proposed: The pre-seismic MHz EM emission is thought to be due to the fracture of the highly heterogeneous system that surrounds the family of large high-strength entities distributed along the fault sustaining the system, while the kHz EM radiation is due to the fracture of the aforementioned large high-strength entities themselves [e.g.,28,29,31,34,39].

The temporal evolution of a MHz EM precursor, which behaves as a phase transition of second order (see Section 3), reveals transition from the phase from non-directional almost symmetrical cracking distribution to a directional localized cracking zone that includes the backbone of strong asperities (*symmetry breaking*). The identification of the time interval where the *symmetry breaking* is completed indicates that the fracture of heterogeneous system in the focal area has been obstructed along the backbone of asperities that sustain the system: *The siege of strong asperities begins*. However, the prepared EQ will occur if and when the local stress exceeds fracture stresses of asperities. As it is mentioned, the lounge of the kHz EM activity shows the fracture of asperities sustaining the fault.

(iii) How can we identify precursory symptoms in EM observations which signify that the occurrence of the prepared EQ is unavoidable?

This is a crucial question. Our results suggest that the appearance of a really seismogenic MHz EM anomaly does not mean that the EQ is unavoidable [28, 29]. The interplay between the heterogeneities and the stress field could be responsible for the observed antipersistent pattern of the precursory MHz EM time series [28, 29]. Indeed, in natural rock at large length scales there are long-range anticorrelations, in the sense that a high value of a rock property, e.g., threshold for breaking is followed by a low value and vice versa. The antipersistent character of the MHz EM time series may reflect the fact that in heterogeneous media, volumes with a low threshold for breaking alternate with much stronger volumes. Crack growth in a heterogeneous medium continues until a much stronger region is encountered. When this happens, crack growth stops while another crack nucleates in a weaker region and so on. Antipersistent behavior implies a set of fluctuations tending to induce stability within the system, i.e., a nonlinear negative feedback, which "kicks" the opening cracks away from extremes. Consequently, heterogeneity could account for the appearance of a stationary-like behavior in the antipersistent MHz part of the prefracture EM time series and thus enable the fracture in highly heterogeneous systems to be described via an analogy with thermal continuous phase transition of second order (see Section 3).

On the contrary, the lounge of the kHz EM activity is the sign of EQ generation. Accumulated evidence support the hypothesis that the kHz EM emission is originated during the fracture of asperities distributed along the activated fault sustaining the system (see Sections 4-7).

The burden of this contribution was to describe a plausible scenario for the study of EM precursors which includes a rather strict set of criteria for characterizing a sequence of MHz - kHz EM emissions as a seismogenic one. We emphasize that this scenario has already been applied to precursors associated with significant, i.e., EQs with magnitude larger than 6, surface EQs that occurred on land or near the coast-line in Greece and Italy. It seems to provide a coherent framework which ties together the observed phenomenology of MHz and kHz EM precursors, without obvious internal inconsistencies and without violating the laws of physics.

It might be difficult for someone to accept that such anomalies are indeed seismogenic. However it is even more difficult to prove that they are not. How possible would it be to find a non seismogenic EM emission that meets the criteria for such a multidisciplinary scheme?

One of the largest controversial issues of the materials science community is the interpretation of scaling laws on material strength. In particular, an important open question is whether the spatial and temporal complexity of earthquake and fault structures emerges from geometry or from the chaotic behaviour inherent to the nonlinear equations governing the dynamics of these phenomena. The observed scaling laws associated with EQs have led a variety of researchers to the conclusion that these events can be regarded as a type of generalized phase transition, similar to the nucleation and critical phenomena that are observed in thermal and magnetic systems [132]. In spite of this prevailing view, other scientists propose a different argument, purely based on geometry. They conclude that as happened for relativity, geometry could again hold an unexpected and fundamental role [133].

Our analysis suggests that we should discriminate two distinct cases: (i) The scaling laws associated with the fracture of the backbone of asperities of a single fault could be a product of the fractal scaling of asperities. Geometry holds a fundamental role of the emergence of fractal scaling laws in phenomena associated with the fracture of asperities. The observed precursory kHz EM emission is such a phenomenon. (ii) The scaling laws associated with the fracture of highly heterogeneous component that surrounds the family of asperities could be attributed to a phase transition of second order. Recent results support the concept that seismicity which preceeds of a significant seismic event is a critical phenomenon, it can be attributed to a phase transition of second order [134]. Moreover, it has been found empirically that main shocks occur a few days up to one week after the appearance of criticality. We recall that the MHz EM precursors also behave as a phase transition of second order, and also emerge from approximately one week up to a few hours before the EQ occurrence. These findings verify that the seismicity and the precursory MHz EM activity are two faces of the same coin. Notice, the persistent kHz EM emission, which is emerged in the tail of the preseismic preseismic EM activity, is a nonequilibrium process without any footprint of an equilibrium thermal phase transition. This process indicates that the system acquires a self-regulating character and to a great degree the property of irreversibility, which is one of the important components of predictive capability. The above mentioned findings suggest reconsidering the interpretation of scaling laws on material strength.

The absence of any EM activity during the EQ occurrence and aftersocks period constitutes a puzzling feature in the study of seismogenic EM precursors. A catastrophic decrease in the elastic modulus just before the final rupture is expected. The appearance of an EM gap in all the frequency bands just before the EQ occurrence might be considered as a hallmark that the expected decrease in the elastic modulus has occurred [28, 29]. So, the existence of a quiescent period may constitute the last clue that a significant seismic event is forthcoming with a considerable probability. On the basis of our study, drawing on both field observations and laboratory experiments on rock fracture, we make the following suggestion concerning the initial and final times for the crucial last stage of the EQ preparation process. The initial point corresponds to the appearance of persistent kHz EM emission. The final point corresponds to the onset of a quiescent period when all precursory EM activities cease. This analysis may point to a possible way of estimating the time to global failure. Certainly, further work in this direction is needed.

Irreversible deformation of rocks is accompanied by the Kaizer effect: if the heterogeneous material is loaded, then unloaded before fracture, and loaded again, only a small number of micro-fractures are detected before attaining the previous load. Micro-fracturing activity increases dramatically as soon as the largest previously experienced stress level are exceeded indicating the beginning of further damage in rocks. The existence of Kaiser effect in geological scale can justify the systematically observed absence of EM emission during the aftershocks period. The stress during the aftershocks period does not exceed the maximum previously reached stress level associated with the main shock occurrence.

The described here results seem to be tolerable, whether the presented ideas will prove to be corrects or disappear as others have remain for the future. However, if we accept the presented suggestions, the absence of EME after the EQ occurrence supports the hypothesis that the launched EQ was the main shock. In any case, the complexity of EQ preparation process is enormous, and thus a huge amount of research is needed before we begin to understand it. There are many outstanding answers that we do not know. Yet it is certain that we have begun to place most of the right questions. And this is perhaps a sign of a latent solution. The Greek poet and Nobel Laureate George Seferis has referred to what the ancient Greek spirit is all about:

"The birthplace of this idea is found at the dawn of Greek history. Aeschylus, the ancient Greek playwright, formulated it once and for all: He who steps beyond moderation is a hubrist, i.e. arrogant, and hubris is the greatest evil that can fall upon us. Greek Tragedy throughout is full of symbols of this idea. And the symbol that moves me above all others, this symbol I find in the Persians. Xerxes, the old legend tells us, was defeated because he was a hubrist; because he committed this extraordinary deed: he lashed at the sea...".

For the purpose of this chapter, it would mean committing hubris for scientists who have dedicated themselves to the prognosis of earthquakes to think that they can defeat "Eggelados".

10. References

[1] Uyeda, S., Nagao, T., and Kamogawa, M.: Short-term earthquake prediction: Current status of seismo-electromagnetics, Tectonophysics 470 205–213, 2009.

[2] Geller, R., Jackson, D., Kagan, Y., Mulargia, F.: Earthquakes cannot be predicted, Science 275, 1616–1617, 1997.

[3] Wyss, M., Martirosyan, A.: Seismic quiescence before the M7, 1988, Spitak earthquake, Armenia. Geophys. J. Int. 134, 329–340, 1998

[4] Huang, Q.: Search for reliable precursors: a case study of the seismic quiescence of 2000 western Tottori prefecture earthquake. J. Geophys. Res. 111, B04301, 2006.

[5] Huang, Q., Sobolev, G.A., Nagao, T.: Characteristics of the seismic quiescence and activation patterns before the M = 7.2 Kobe earthquake, January 17, 1995, Tectonophysics 237, 99–116, 2001.

[6] Kossobokov, V.G., Romashkova, L.L., Keilis-Borok, V.I., Healy, J.H.: Testing earthquake prediction algorithms: statistically significant real-time prediction of the largest earthquakes in the Circum-Pacific, 1992–1997, Phys. Earth Planet. Inter. 111, 187–196,

[7] Keilis-Borok, V., Shebalin, P., Gabrielov, A., Turcotte, D.: Reverse detection of short term earthquake precursors. Phys. Earth Planet. Inter. 145, 75–85, 2004.

[8] Bahat, D., Rabinovitch, A., and Frid, V.: Tensile Fracturing in Rocks , Springer, New York, 2005.

[9] Ogawa, T., Oike, K. and Miura, T.; Electromagnetic radiation from rocks. J. Geophys. Res. 90, 6245–6249, 1985.

[10] OŠKeefe, S. G. and Thiel, D. V.; A mechanism for the production of electromagnetic radiation during fracture of brittle materials. Phys. Earth Plant. Inter. 89, 127–135, 1995.

[11] Lolajicek, T. and Sikula, J.: Acoustic emission and electromagnetic effects in rocks. In: Progress in Acoustic Emission VIII. Proceedings of the 13th International Acoustic Emission Symposium, 30 November, 1996. (Kishi, T., Mori, Y., Higo, H. and Enoki, M., Eds). Japanese Society for NDI, Nara, Japan: 311–314: 1996.

[12] Panin, V., Deryugin, Ye., Hadjicontis, V., Mavromatou, C., and Eftaxias, K.: Scale levels of strain localization and fracture mechanism of LiF single crystals under compression, Physical Mesomechanics, 4, 21-32, 2001.

[13] Frid, V., Rabinovitch, A. and Bahat, D.: Fracture induced electromagnetic radiation. J. Phys. D. Appl. Phys. 36, 1620–1628, 2003.

[14] Mavromatou, C., Hadjicontis, V., Ninos, D. Mastroyiannis, D., Hadjicontis, E., and Eftaxias, K.: Understanding the fracture phenomena in inhomogeneous rock samples and ionic crystals, by monitoring the electromagnetic emission during the deformation, Physics and Chemistry of the Earth, 29, 353 – 357, 2004.

[15] Fukui, K., Okubo, S. and Terashima, T.: Electromagnetic radiation from rock during uniaxial compression testing: the effects of rock characteristics and test conditions. Rock Mech. Rock Eng. 38, 411–423, 2005.

[16] Lacidogna, G., Carpinteri, A., Manuello, A., Durin, G., Sciavi, A., Niccolini, G., and Agosto, A.: Acoustic and electromagnetic emissions as precursor phenomena in failure processes, Strain 47,1-9, 2011, doi: 10.1111/j.1475-1305.2010.00750.x

[17] Warwick, J. W., Stoker. C, and Meyer, T. R.: Radio emission associated with rock fracture: possible application to the great Cjilean earthquake of May 22, 1960, J. Geophys. Res. 87, 2851-2859, 1982.

[18] Gokhberg, M. B., Morgunov, V. A., Yoshino, T. and Tozawa, I.: Experimental measurement of electromagnetic emissions possibly related to earthquakes in Japan. J. Geophys. Res. 87, 7824–7828, 1982

[19] Hayakawa, M. and Fujinawa, Y.: Electromagnetic Phenomena Related to Earthquake Prediction, Terrapub, Tokyo, 1994.

[20] Hayakawa, M.: Atmospheric and Ionospheric Electromagnetic Phenomena Associated with Earthquakes, Terrapub, Tokyo, 1999.

[21] Eftaxias, K., Kopanas, J., Bogris, N., Kapiris, K., Antonopoulos, G. and Varotsos P.: Detection of electromagnetic earthquake precursory signals in Greece, Proc. Japan Acad., 76(B), 45-50, 2000.

[22] Eftaxias, K., P. Kapiris, J. Polygiannakis, N. Bogris, J. Kopanas, G. Antonopoulos, A. Peratzakis and V. Hadjicontis.: Signatures of pending earthquake from electromagnetic anomalies. Geophys. Res. Let., 28, 3321-3324, 2001.

[23] Hayakawa, M. and Molchanov, O.: Seismo Electromagnetics, Terrapub, Tokyo, 2002.

[24] Nagao, T., Enomoto, Y., Fujinawa, Y. etäal.: Electromagnetic anomalies associated with 1995 Kobe earthquake. J. Geodyn. 33, 401–411, 2002.

[25] Eftaxias, K., Kapiris, P., Dologlou, E., Kopanas, J., Bogris, N., Antonopoulos, G., Peratzakis, A., and Hadjicontis, V.: EM anomalies before the Kozani earthquake: A study of their behaviour through laboratory experiments, Geophys. Res. Lett., 29, 69/1-69/4, 2002.

[26] Eftaxias, K., Kapiris, P., Polygiannakis, J., Kopanas, J., Antonopoulos, G., and Rigas, D.: Experience of short term earthquake precursors with VLF-VHF electromagnetic emissions, Natural Hazards and Earth System Sciences, 3, 217-228, 2003.

[27] Eftaxias, K., Frangos, P., Kapiris, P., Polygiannakis, J., Kopanas, J., Peratzakis, A., Skountzos, P., and Jaggard, D.: Review-Model of Pre-Seismic Electromagnetic Emissions in Terms of Fractal-Electrodynamics, Fractals, 12, 243 – 273, 2004.

[28] Kapiris, P., Eftaxias, K., Chelidze, T.: Electromagnetic Signature of Prefracture Criticality in Heterogeneous Media, Physical Review Letters, 92(6), 065702, 2004.

[29] Contoyiannis, Kapiris, P., and Eftaxias, K.: A Monitoring of a Pre-Seismic Phase from its Electromagnetic Precursors, Physical Review E, 71, 061123-1 – 061123-14, 2005.

[30] Karamanos, K., Dakopoulos, D., Aloupis, K., Peratzakis, A., Athanasopoulou, L., Nikolopoulos, S., Kapiris, P., Eftaxias, K.: Study of pre-seismic electromagnetic signals in terms of complexity. Physical Review E. 74, 016104-1/21, 2006.

[31] Eftaxias, K., Sgrigna, V., and Chelidze, T., (Eds): Mechanical and Electromagnetic Phenomena Accompanying Preseismic Deformation: from Laboratory to Geophysical Scale, Tectonophysics, 431, 1-301, 2007.

[32] Papadimitriou, K., Kalimeri, m., and Eftaxias, K.: Nonextensivity and universality in the earthquake preparation process, Physical Review E, 77, 36101, 2008.

[33] Kalimeri, M., Papadimitriou, K., Balasis, G., and Eftaxias, K.: Dynamical complexity detection in pre-seismic emissions using nonadditive Tsallis entropy, Physica A, 387, 1161-1172-, 2008.

[34] Contoyiannis, Y., and Eftaxias, K.: Tsallis and Levy statistics in the preparation of an earthquake, Nonlinear Processes in Geophysics, 15, 379–388, 2008.

[35] Eftaxias, K., Athanasopoulou, L., Balasis, G., Kalimeri, M., Nikolopoulos, S., Contoyiannis, Y., Kopanas, J., Antonopoulos, G., and Nomicos, C.: Unfolding the procedure of characterizing recorded ultra low frequency, kHZ and MHz electromagetic anomalies prior to the L'Aquila earthquake as pre-seismic ones. Part I, Nat. Hazards Earth Syst. Sci.., 9, 1953–1971, 2009.

[36] Eftaxias, K., Balasis, G., Contoyiannis, Y., Papadimitriou, C., Kalimeri, M., Kopanas, J., Antonopoulos, G., and Nomicos, C.: Unfolding the procedure of characterizing recorded ultra low frequency, kHZ and MHz electromagnetic anomalies prior to the L'Aquila earthquake as pre-seismic ones. Part II, Nat. Hazards Earth Syst. Sci. 10, 275–294, 2010.

[37] Eftaxias, K., Maggipinto, T., Meister, C-V., and Katz. O (Eds).: Progress in the research on earthquakes precursors, Natural Hazard and Earth System Sciences (Special Issue), 2011.

[38] Kossobokov, V.: Testing earthquake prediction methods: the West Pacific short-term forecast of earthquakes with magnitude $MwHRV > 5.8$, Tectonophysics, 413, 25–31, 2006

[39] Contoyiannis, Y., Nomicos, C., Kopanas, J., Antonopoulos, G., Contoyianni , L.,and Eftaxias, K.: Critical features in electromagnetic anomalies detected prior to the L'Aquila earthquake, Physica A 389 , 499-508, 2010.

[40] Herrmann, H. J., and Roux, S.: Statistical Physics for the Fracture of Disordered Media, Elsevier, Amsterdam, 1990.

[41] Sornette, D.: Critical Phenomena in Natural Sciences, Chaos, Fractals, Self-organization and Disorder: Concepts and Tools, Second edition, Springer Series in Synergetics, Heidelberg, 2004.

[42] Contoyiannis, Y., Diakonos, F., Kapiris, P., Peratzakis, A., and Eftaxias, K.: Intermittent Dynamics of Critical Pre-seismic Electromagnetic Fluctuations, Physics and Chemistry of the Earth, 29, 397 – 408, 2004.

[43] Mandelbrot, B.: The Fractal Geometry of Nature, W. H. Freema, New York, 1982.

[44] Jaggard, D.: On fractal electrodynamics, in Recent Advances in Electromagnetic Theory, eds. H. Kritikos and D. Jaggard, Springer-Verlag, New York, 183–224, 1990.

[45] Jaggard, D., and Frangos, P.: Surfaces and superlattices, in Frontiers in Electrodynamics, eds. D. Werner and R. Mittra, IEEE Press,1–47, 2000.

[46] Varotsos, P.: The Physics of Seismic Electric Signals, TerraPub, Tokyo, 2005.

[47] Pulinets, S. and Boyarchuk, K.: Ionospheric Precursors of Earthquakes, Springer, 2005.

[48] Ouzounov, D., and Freund, F.: Mid-infrared emission prior to strong earthquakes analyzed by remote sensing data. Advances in Space Research, 33, 268–273, 2004.

[49] Rosen, P., Hensley, S., Joughin, I., Li, F., Madsen, S., Rodriguez, E., and Goldstein, R.: Synthetic Aperture Radar Interferometry, Proceedings of the IEEE, 88, 333-382, 2000

[50] Stanley, H.: Scaling, universality, and renormalization: Three pillars of modern critical phenomena, Rev. Mod. Phys., 71, S358– S366, 1999.

[51] Bar-Yam, Y.: Dynamics of complex systems. Reading, Mass., Addison-Wesley, 1997.

[52] Contoyiannis, Y. and Diakonos, F.: Criticality and intermittency in the order parameter space, Phys. Lett. A, 268, 286–272, 2000.

[53] Contoyiannis, Y., Diakonos, F., and Malakis, A.: Intermittent dynamics of critical fluctuations, Phys. Rev. Lett., 89, 35701– 35704, 2002.

[54] Varotsos, P., Sarlis, N., Skordas, E., Uyeda, S., and Kamogawa, M.: Natural time analysis of critical phenomena, PNAS, July 12, 108 , 11361–11364, 2011.

[55] Titchener, M., Nicolescu, R., Staiger, L., Gulliver, A., and Speidel,U.: Deterministic Complexity and Entropy, Fund. Inform., 64, 443–461, 2005.

[56] Fisher, R.: Theory of statistical estimation, Proc. Camb. Phil. Soc. 22, 700–725, 1925.

[57] Grassberger, P. and Procaccia, I.: Characterization of strange attractors, Phys. Rev. Lett., 50, 346–349, 1983.

[58] Hurst, H.: Long term storage capacity of reservoirs, Trans. Am. Soc. Civ. Eng., 116, 770–808, 1951.

[59] Peng, C., Mietus, J., Hausdorff, J., Havlin, S., Stanley, H., and Goldberger, A.: Long-range anticorrelations and non-Gaussian behavior of the heartbeat, Phys. Rev. Lett., 70, 1343–1346, 1993.

[60] Peng, C., Havlin, S., Stanley, H., and Goldberger, A.: Quantification of scaling exponents and crossover phenomena in nonstationary heartbeat timeseries, Chaos, 5, 82-87, 1995.

[61] Shannon, C. E.: A mathematical theory of communication, The Bell System Tech. J., 27, 379–423, 623-656, 1948.

[62] Ebeling, W. and Nicolis, G.: Word frequency and entropy of symbolic sequences: A dynamical Perspective, Chaos, Solitons & Fractals, 2, 635–650, 1992.

[63] Ebeling, W.: Prediction and entropy of nonlinear dynamical systems and symbolic sequences with LRO, Physica D, 109, 42–52, 1997.

[64] Ebeling, W., Steuer, R., and Titchener, M.: Partition-based entropies of deterministic and stochastic maps, Stochastics and Dynamics, 1, 45–61, 2001.

[65] Ebeling,W.: Entropies and predictability of nonlinear processes and time series, edited by: Sloot, P. M. A., et al., ICCS 2002, LNCS, 1209–1217, 2002

[66] Tsallis, C.: Possible generalization of Boltzmann-Gibbs statistics, J. Stat. Phys., 52, 479–487, 1988.

[67] Tsallis, C.: Introduction to Nonextensive Statistical Mechanics, Approaching a Complex Word, Springer, 2009.

[68] Karamanos, K. Peratzakis, A., Kapiris, P., Nikolopoulos, S., Kopanas, J., and Eftaxias, K.: Extracting pre-seismic electromagnetic signatures in terms of symbolic dynamics, Nonlinear Processes in Geophysics, 12, 835-848, 2005.

[69] Nikolopoulos, S., Kapiris, P., Karamanos K., and Eftaxias, K.: A unified approach of catastrophic events, Natural Hazards and Earth System Sciences, 4, 615-637, 2004

[70] Eftaxias, K., Kapiris, P., Balasis, G., Peratzakis, A., Karamanos, K., Kopanas, J., Antonopoulos, G., and Nomicos, C.: A Unified Approach to Catastrophic Events: From the Normal State to Geological or Biological Shock in Terms of Spectral Fractal and Nonlinear Analysis, Natural Hazards and Earth System Sciences, 6, 205-228, 2006.

[71] Eftaxias, K., Minadakis, G., Athanasopoulou. L., Kalimeri. M., Potirakis, S., and Balasis, G.: Are Epileptic Seizures Quakes of the Brain? An Approach by Means of Nonextensive Tsallis Statistics (submitted)

[72] Sotolongo-Costa, O. and Posadas, A.: Fragment-asperity interaction model for EQ, Phys. Rev. Lett., 92, 048501, 2004.

[73] Silva, R., Franca, G., Vilar, C., and Alcaniz, J.: Nonextensive models for earthquakes, Phys. Rev. E, 73, 026102, 1–5, 2006.

[74] Telesca, L.: Tsallis-Based Nonextensive Analysis of the Southern California Seismicity Entropy, 13(7), 1267-1280, 2011.

[75] Eftaxias, K.: Footprints of nonextensive Tsallis statistics, self-affinity and universality in the preparation of the L'Aquila earthquake hidden in a pre-seismic EM emission, Physica A 389, 133-140, 2009.

[76] Huang, J., and Turcotte, D.: Fractal distributions of stress and strength and variations of b–value, Earth Planet. Sci. Lett., 91, 223-230, 1988.

[77] Rabinovitch, A., Frid, V., and Bahat, D.: Gutenberg-Richter-type relation for laboratory fracture-induced electromagnetic radiation, Phys. Rev. E, 65, 11 401/1-11 401/4, 2001.

[78] Kapiris, P., Balasis, G., Kopanas, J., Antonopoulos, G., Peratzakis, A., and Eftaxias, K.: Scaling Similarities of Multiple Fracturing of Solid Materials, Nonlinear Proc. Geoph., 11, 137–151, 2004.

[79] Scholz, C.: The frequency-magnitude relation of macrofracturing in rocks and its relation to earthquakes, Bull. Seismo. Soc. Am., 58, 399–415, 1968.

[80] Ponomarev, A. Zavyalov, A., Smirnov, V., and Lockner, D.: Physical modelling of the formation and evolution of seismically active fault zones, Tectonophysics, 277, 57–81, 1997.

[81] Lei, X., and Satoh, T.: Indicators of critical point behavior prior to rock failure inferred from pre-failure damage, Tectonophysics, 431, 97–111, 2007.

[82] Hainzl, S., Zoller, G., and Scherbaum, F.: Earthquake clusters resulting from delayed rupture propagation in finite fault segments, Geophys. Res. Lett., 108, 2013-2016, 2003.

[83] Heneghan C., and McDarby, G.: Establishing the relation between detrended fluctuation analysis and power spectral density analysis for stochastic processes, Phys. Rev. E, 62, 6103–6110, 2000

[84] Turcotte., D.: Fractals and chaos in geology and geophysics, Cambridge University Press, 1997.

[85] Sammis, C. and Sornette, D.: Positive feedback, memory, and the predictability of EQ, P. Natl. Acad. Sci. USA, 99, 2501–2508, 2002.

[86] Ponson, L., Bonamy, D., and Bouchaud, E.: Two-dimensional scaling properties of experimental fracture surfaces, Phys. Rev. Lett., 96, 35506-1/4, 2006.

[87] Mourot, G., Morel, S., Bouchaud, E., and Valentin, G.: Scaling properties of mortar fracture surfaces, Int. J. of Fracture, 140, 39–54, 2006.

[88] Lopez, J., and Schmittbuhl, J.: Anomalous scaling of fracture surfaces, Phys. Rev. E, 57, 6405-6408, 1998.

[89] Zapperi, S., Kumar, P., Nukala, V., and Simunovic, S.: Crack roughness and avalanche precursors in the random fuse model, Phys. Rev. E, 71, 26106/1–10, 2005.

[90] Hansen, A., and Schmittbuhl, J.: Origin of the universal roughness exponent of brittle fracture surfaces:stress-weighted percolation in the damage zone, Phys. Rev. Lett., 90, 45504–45507, 2003.

[91] Renard, F., Voisin, C., Marsan, D., and Schmittbuhl, J.: High resolution 3D laser scanner measurements of a strike-slip fault quantity its morphological anisotropy at all scales, Geophys. Res. Lett., 33, L04305, 2006.

[92] Sornette, D., Helmstetter, A.: Occurrence of finite-time singularities in epidemic models of rupture, earthquakes and starquakes, Phys. Rev. Lett. 89 (15), 158501, 2002.

[93] Sornette, D., Sammis, C.: Complex critical exponents from renormalization group theory of earthquakes: Implications for earthquake predictions, J. Phys. I 5, 607–619, 1995.

[94] Sornette, D., Vanneste, C.: Dynamics and memory effects in rupture of thermal fuse networks, Phys. Rev. Lett. 68, 612–615, 1992.

[95] Sornette, D., Vanneste, C., Knopoff, L.: Statistical model of earthquake foreshocks, Phys. Rev. A 45, 8351–8357.

[96] Bowman, D., Ouillon, G., Sammis, C., Sornette, A., and Sornette, D.: An observational test of the critical earthquake concept, J. Geophys. Res., 103, 24359-24372, 1998.

[97] Bowman, D. and King, G.: Accelerating seismicity and stress accumulation before large Earthquakes, J. Geophys. Res. Lett., 28(21), 4039–4042, 2001.

[98] Bufe, C. and Varnes, D.: Predictive modelling of the seismic cycle of the greater San Francisco Bay region, J. Geophys. Res., 98, 9871–9883, 1993.

[99] S. C. Jaume and L. R. Sykes, Evolving towards a critical point: a review of accelerating seismic moment/energy release prior to large and great earthquakes, Pure Appl. Geophys. 115 (1999) 279–305.

[100] Sahimi, M.: Flow phenomena in rocks: from continuum models to fractals, percolation, cellular automata, and simulated annealing, Rev. Mod. Phys., 65, 1393–1534, 1993.

[101] Sahimi, M., Robertson, M., and Sammis, C.: Fractal distribution of earthquakes hypocenters and its relation to fault patterns and percolation, Phys. Rev. Lett., 70, 2186–2189, 1993.

[102] Sornette, D.: Self-organized criticality in plate tectonics, in: Spontaneous Formation of Space-Time Sructures and Criticality, edited by Riste, T. and Sherrington, D., 57–106, Kluwer Academic Publishers, 1991.

[103] Varotsos, P.: The Physics of Seismic Electric Signals, TerraPub, Tokyo, 2005.

[104] Kapiris, P., Nomicos, K., Antonopoulos, G., Polygiannakis, J., Karamanos, K., Kopanas, J., Zissos, A., Peratzakis, A., and Eftaxias, K.: Distinguished seismological and electromagnetic features of the impending global failure: did the 7/9/1999 M5.9 Athens earthquake come with a warning? Earth Planets and Space, 57, 215-230, 2005.

[105] Potirakis, S., Minadakis, G., Eftaxias, K.: Fisher information measure, Tsallis entropy, Symbolic dynamics, Fracture induced electromagnetic emissions, Physica A, (in press).

[106] Pulinets, S., LegenŠka, A., Gaivoronskaya, T., and Depuev, V: Main phenomenological of ionospheric precursors of strong earthquakes, J. Atmos. Sol.-Terr. Phy., 65, 1337–1347, 2003.

[107] Vicsek, T.: A question of scale, Nature, 411, 421 pp., 2001.

[108] Vicsek, T.: The bigger picture, Nature, 418, 131 pp., 2002.

[109] Stanley, H.: Exotic statistical physics: Applications to biology, medicine, and economics, Physica A, 285, 1-17, 2000.

[110] Picoli, S., Mendes, R., Malacarne, L., Papa, A.: Similarities between the dynamics of geomagnetic signal and heartbeat intervals, Europhysics Letters, 80, 50006/1Ŭ6, 2007.

[111] de Arcangelis, L., Godano, C., Lippiello, E., and Nicodemi, M.: Universality in Solar Flare and Earthquake Occurrence, Phys. Rev. Lett., 96, 051102/1–4, 2006.

[112] Kossobokov, V., Keillis-Borok, V., and Cheng, B.: Similarities of multiple fracturing on a neutron star and on Earth, Phys. Rev. E, 61, 3529–3533, 2000.

[113] Sornette, D.: Predictability of catastrophic events: material rupture, earthquakes, turbulence, financial crashes and human birth, Proceedings of the National Academy of Sciences USA, 99, 2522–2529, 2002.

[114] Abe, S., and Suzuki, N.: Statistical similarities between internetquakes and earthquakes, Physica D 193, 310-314, 2004.

[115] Peters, O., Hertlein, C., and Christensen, K.: A complexity view of rainfall, Phys. Rev. Lett. 88, 018701, 2002.

[116] Fukuda, K., Nunes, L., and Stanley, H.: Similarities between communication dynamics in the Internet and the automatic nervous system, Europhys. Lett., 62, 189–195, 2003.

[117] Herz, A. and Hopfield, J.: Earthquake cycles and neural reverberations: Collective oscillations in systems with pulse-coupled threshold elements, Phys. Rev. Lett., 75, 1222-1225, 1995.

[118] Hopfield, J.: Neurons, dynamics and computation, Phys. Today, 40, 40-46, 1994.

[119] Usher, M., Stemmler, M., and Olami, Z.: Dynamic pattern formation leads to $1/f$ noise in neural populations, Phys. Rev. Lett., 74, 326–329, 1995.

[120] Corral, A., Perez, C., and Diaz-Guilera, A.: Self-organized criticality induced by diversity, Phys. Rev. Lett., 78(8), 1492–1495, 1997.

[121] Plenz, D.: When inhibition goes incognito: feedback interaction between spiny projection neurons in striatal function, TRENDS in Neurosciences, 26(8), 436–443, 2003.

[122] Zhao, X. and Chen, T.: Type of self-organized criticality model based on neural networks, Phys. Rev. E, 65, 026114-1–026114-6, 2002.

[123] Rundle, J., Tiampo, K., Klein, W., and Sa Martins, J.: Selforganization in leaky threshold systems: the influence of near mean field dynamics and its implications for EQs, neurology, and forecasting, PNAS, 99, 2514–2521, 2002.

[124] Nikolopoulos, S., Kapiris, P., Karamanos, K., and Eftaxias, K.: A unified approach of catastrophic events, Natural Hazards and Earth System Sciences, 4, 615-637, 2004.

[125] Li, X., Polygiannakis, J., Kapiris, P., Peratzakis, A., Eftaxias, K., and Yao, X.: Fractal spectral analysis of pre-epileptic seizures in terms of criticality, Journal of Neural Engineering 2, 1-6, 2005.

[126] Kapiris, P., Polygiannakis, J., Yao, X., and Eftaxias, K.: Similarities in precursory features in seismic shocks and epileptic seizures. Europhysics Letters 69, 657-663, 2005.

[127] Osorio, I., Frei, M., Sornette, D., Milton, J., and Lai, Y.: Epileptic seizures: Quakes of the brain? Phys. Rev. E. 82, 021919, 2010.

[128] Balasis, G., Daglis, I., Kapiris, P., Mandea, M., Vassiliadis, D., and Eftaxias, K.: From pre-storm activity to magnetic storms: a transition described in terms of fractal dynamics, Ann. Geophys. 24, 3557-3567, 2006.

[129] Balasis, I. Daglis, Papadimitriou, C., Kalimeri, M., Anastasiadis, A., and Eftaxias, K.: Investigating dynamical complexity in the magnetosphere using various entropy measures, Journal of Geophysical Research, 2009.

[130] Balasis, G., Daglis, I., Papadimitriou, C., Kalimeri, M., Anastasiadis, A., and Eftaxias, K.: Dynamical complexity in Dst time series using non-extensive Tsallis entropy, Geophysical Research Letters, L14102, doi:10.1029/2008GL034743, 2008.

[131] Balasis, G., Daglis, I., Anastasiadis, A., Papadimitriou, C., Mandea, M., and Eftaxias, K.: Universality in solar flare, magnetic storm and earthquake dynamics using Tsallis statistical mechanics, Physica A, 390, 341-346, 2011.

[132] Rundle, J., Turcotte, D., Shcherbakov, R., Klein, W., and Sammis, C.: Statistical physics approach to understanding the multiscale dynamics of earthquake fault systems, Reviews of Geiophysics, 41, 5/1-5/30, doi:10.1029/2003RG000135, 2003.

[133] Carpinteri, A., and Pugno, N.: Are scaling laws on strength of solids related to mechanics or to geometry?, Nature materials, 4, 421-423, 2005.

[134] Varotsos, P., Sarlis, N., and Skordas, E.: Natural Time Analysis: The New View of Time (Springer, Berlin), 2011.

11

Changes in Apparent Resistivity in the Late Preparation Stages of Strong Earthquakes

Du Xuebin et al.*
Lanzhou Base of Institute of Earthquake Science, CEA, Lanzhou
China

1. Introduction

In China, a large-scale observation network that is comprised of a number of apparent resistivity (for short, AR) stations has been established for the purpose of earthquake (EQ) monitoring and prediction since the 1966 Ms7.2 Xingtai EQ in Hebei Province. Presently, over 70 AR stations are in observation in seismically active belts in densely populated areas and nearby some of large, medium-sized cities. The 2008 Ms8.0 Wenchuan great EQ in Sichuan province occurred in the AR station network that was located in the border area of both Sichuan and Gansu provinces. In the 1970s-1980s, more than 110 AR stations had been in observation, and in those years, several great earthquakes (EQs) occurred nearby AR stations, such as the 1976 Ms7.8 Tangshan EQ in Hebei province, the 1976 Ms7.2 Songpan-Pingwu EQ in Sichuan province, the 1976 Ms7.4 Longling EQ in Yunnan province, and the 1988 Ms7.6 Lancang-Gengma EQ in Yunnan Province. At an AR station, two horizontally perpendicular observation channels or three channels, more one horizontally skewed channel (a NE or NW channel, as illustrated in Fig.1a), are employed, and for each channel, an AR observation configuration with a symmetry four-electrode array is installed (Fig.1b). For most stations, the current electrode spacing $\overline{AB} = 1000 \sim 2000\text{m}$.

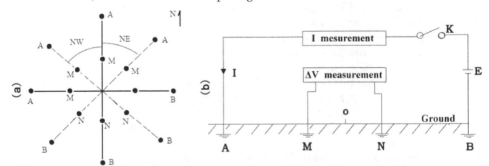

Fig. 1. The observation channels (a) and symmetric four-electrode resistivity array for a channel (b) at a geo-electrical station

* An Zhanghui[1], Yan Rui[3], Ye Qing[2], Fan Yingying[1], Liu Jun[1], Chen Junying[1] and Tan Dacheng[1]
[1]*Lanzhou Base of Institute of Earthquake Science, CEA, Lanzhou, China*
[2]*China Earthquake Networks Center, China Earthquake Administration, Beijing, China*
[3]*Earthquake Administration of Beijing Municipality, Beijing, China*

Some of the AR stations have continuously observed for more than 40 years, and a lot of AR data are observed. Using these data we can understand whether the earthquake-related AR changes are recorded or not, which is an issue that should be seriously discussed because the precursory anomalies before EQs have been strongly debated. In this chapter we try to study the issue from two respects: (1) the EQ case research on AR changes recorded before EQs, and (2) the theoretical analysis on anisotropic AR changes related to the maximum principal compression stress (P – axis) azimuth of focal mechanism solution, nearby an EQ focal region in the late preparation stages of a strong EQ. This chapter will help objectively evaluate and comprehend the AR precursory changes related to EQs.

2. EQ case research

Before several great EQs with magnitude of Ms≥7.0 and some moderate EQs, obvious AR anomalies, which include the medium-term anomalies that start to appear about 2-3 years to several months before EQs and the imminent anomalies that start to appear about 3 months to several days preceding EQs, are recorded at geo-electrical stations nearby the epicentral areas, in China. Some of the anomalies are ascertained after EQs, whereas some are discerned before EQs. More interestingly, two groups of EQs with magnitude of Ms 6 nearby AR stations are actually forecasted on a one-year time scale using medium-term AR anomalies, for which the expected EQ magnitudes and forecasted locations are all right.

2.1 AR changes ascertained after EQs
2.1.1 Reappearing AR anomalies before two great EQs
In 1976, three great EQs with magnitude of Ms ≥ 7.2 occurred in the mainland Chinese, such as the July 28 Ms7.8 Tangshan EQ in Hebei province, the Aug. 16 Ms7.2 Songpan-Pingwu EQs in Sichuan province, and the May 29 Ms7.4/7.3 Longling EQ in Yunnan province. Of the EQs, the Tangshan EQ and Songpan-Pingwu EQ occurred in a local AR station network, and significant AR changes were ascertained after the EQs. As shown in figure 2[1], obviously drop AR changes were recorded at station Changli-Houtuqiao (CLH) in Hebei province, before the Ms7.8 Tangshan EQ (80 km from CLH station) and at station Wudu-Hanwang(WDH) in Gansu province, before the Songpan-Pingwu Ms7.2 EQ (105 km from WDH station), respectively. On those days, the two stations were in normal operation. It can be seen from raw AR daily mean curves of the two stations that AR changes fell all during about 40 days before the occurrence of the two great EQs, which were notable short-term anomalies proceeding the two EQs. Especially, during about 20 days before the two EQs AR changes started to fall by a larger margin, which were imminent anomalies before the impending EQs. Then, immediately after the occurrence of the two EQs, the drop changes started to rise. Based on the two AR anomalies corresponding to the two EQs, we can discuss two problems as follows:
The first problem is on the repeatability of AR anomalies before the two EQs. The Ms7.8 Tangshan EQ occurred in Hebei Province, in east China, and the M7.2 Songpan-Pingwu EQ occurred in north Sichuan province, in western China. The distance between the two EQ epicenters were beyond 1500 km; and the two stations were located in different tectonic units (station CLH was nearby the Cangdong fault belt in the Beijing-Tianjin-Tangshan area, in east China, and station WDH was nearby the Bailong river fault belt in South Gansu province, in western China). The underground geo-electrical structure of the two stations was very different. Nevertheless, the drop AR changes of the two stations before the two

great EQs and their recovery AR changes immediately after the two events had a similar changeable pattern in appearance. Therefore, the anomalous AR changes were believed to be the anomalies related to the two great EQs.

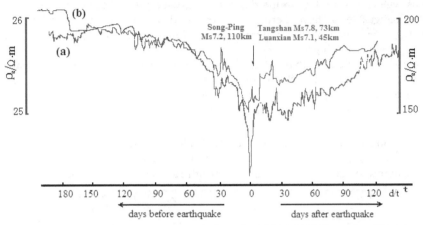

Fig. 2. AR changes observed at CLH station before the 1976, Ms7.8 Tangshan EQ (a), and at WDH station before the 1976, Ms7.2 Songpan-Pingwu EQ (b)

The second problem is on the drop pattern of the two AR anomalies. It is obvious from figure 2 that the AR changes have a notable drop pattern during the imminent stage of the two impending great EQs. In general, the underground medium is abundant with water and the medium resistivity is susceptible to water, therefore, we believed that the changes of underground water resulted in these drop AR changes before the two impending EQs. Figure 3 are the raw AR daily-mean curve at CLH station and the raw curve of water level

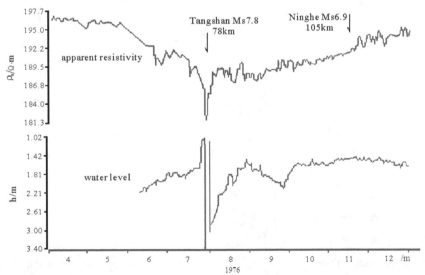

Fig. 3. The AR change observed at CLH station and the water level change observed at Longjiadian station

observed at underground water station[2], Longjiadian station which was about 20 km from station CLH and nearby the Cangdong rupture zone. We can notice from figure 3 that immediately before and after the occurrence date of the EQ, the drop AR change at station CLH was well corresponding to the rise change in water level at station Longjiadian. The opposite changeable patterns between electric and water are quite significant, which indicated that the water in the underground medium nearby the focal region played an important role in AR changes.

It can be seen from figure 3 that nothing was recorded before the Ms6.9 Ninghe aftershock (Nov. 15, 1976) at the geo-electrical and water stations, a possible reason for which was explained by associating with the mainly active faults in/nearby the focal region and the focal mechanism of the aftershock by Du et al.[3-4].

2.1.2 AR anomalies corresponding to the Ms8.0 Wenchuan EQ

On May 12, 2008, a great EQ of Ms8.0 struck the Wenchuan county and its adjacent area in Sichuan province of China, and within one month following it, more than thirty Ms5.0~6.4

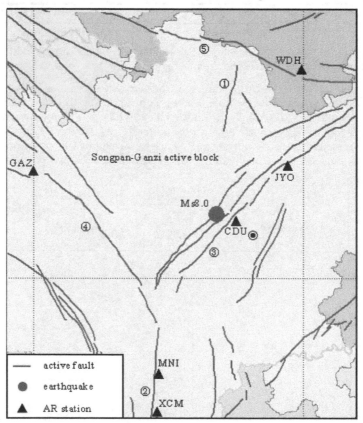

①Minjiang river fault, ②Anning river fault, ③Doujiang Weir–An county fault, ④Xianshui river fault, ⑤White Dragon River fault

Fig. 4. Distributions of station, epicenter and fault.

aftershocks occurred along the NE-strike fault belt beyond 300 km long, from the Wenchuan county to the Ningqiang county in Shaanxi province. There were six AR stations within 400 around the main epicenter, which were Chengdu station (CDU, 35km), Jiangyou station (JYO, 150 km), Ganzi station (GAZ, 331 km), Mianning station (MNI, 260) and Xichang station (XCM, 360 km) in Sichuan province, and WDH station (300 km) in Gansu province (Fig.4).

2.1.2.1 Medium-term AR anomalies before the EQ

Significant AR anomalies were recorded at four stations in the medium-term stage before the Ms 8.0 Wenchuan EQ as follows.

1. Locally concentrated anomalies

The anomalies were recorded at CDU, JYO, GAZ and WDH during the medium-term stage before the EQ, which were in the range of 400km from the main epicenter (Fig.5, Fig.6).

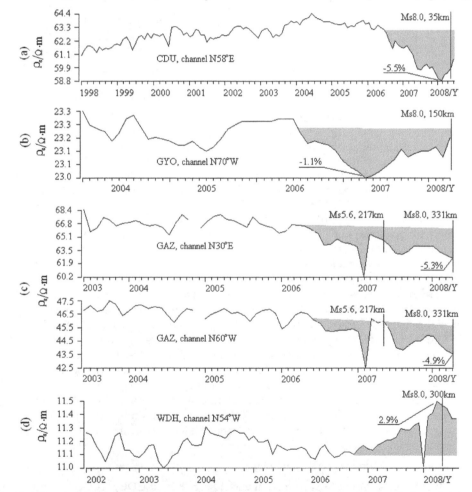

Fig. 5. AR monthly mean changes observed at four stations that were located along the Songpan-Ganzi active block before the M8.0 Wenchuan EQ

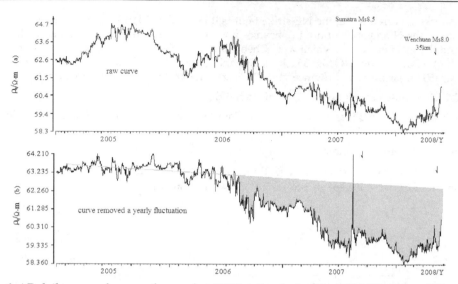

Fig. 6. AR daily mean changes observed at CDU station before the M8.0 Wenchuan EQ

Furthermore, the spatial distribution of the anomalies was tectonically relevant to the Songpan-Ganzi active block. The main shock and its aftershocks occurred along the Longmen mountain nappe structure of the block. Accordingly, the anomalies were recorded at the four stations along the bordering faults around the block (Fig.4), whereas such anomalies were not recorded at the two stations MNI and XCM, which were located along the Anning river fault, beyond the block. This situation is similar to that which was observed before the 1976 Ms7.8 Tangshan EQ when anomalies appeared mostly at the stations along NE- and NW-striking conjugate faults in the Beijing-Tianjin-Tangshan areas.

According to previous statistical studies on numerous EQ cases[5-6] by other Chinese scholars, the spatially distribution of medium-term AR anomalies is about 300~400 km before an EQ with magnitude of Ms ≥ 7.0. According to paper[7], for Ms>6.0 EQs most AR anomalies distribute in the range of 400 km from epicenters and there is commonly no obvious difference between the ranges for Ms>6.0 EQs and Ms ≥ 7.0 EQs. The results show that an epicentral distance of 400km can be used as a reference for identifying medium-term AR anomalies related to Ms>6.0 EQs. Du et al.[8] studied the relationship between the spatial distribution of medium-term anomalies of other precursor observations in China, such as AR, groundwater chemical components and water level, geo-stress and geo-deformation, and the mainly active fault around the epicentral region, as a result, it was believed that the anomalies usually appeared nearby the faults around EQ focal areas.

The spatial distribution of the medium-term AR anomalies before the Wenchuan great EQ, which were concentrated in the range of 400 km from the main epicenter and along the bordering faults around the Songpan-Ganzi active block, was in accordance with the foregoing research results.

2. Synchronous medium-term AR anomalies

These AR anomalies started to appear from about Aug. 2006 at CDU, JYO, GAZ and WDH. In other wards, they appeared synchronously before the main shock at the four stations which were located along the bordering faults of the Songpan-Ganzi active block. The behavior was similar to previous research results. According to Du et al.[8], the medium-term AR anomalies along a actively geological structure around an EQ focal area usually started to appear synchronously or quasi-synchronously. In fact, the medium-term anomalies in observation of groundwater chemical components and water level, geo-stress and geo-deformation along a same structure also have such behavior as seen in AR observation[8].

3. Mostly drop-type AR anomalies

The drop-type anomalies were recorded at stations CDU, GAZ and JYO. Only at station WDH, was a rise-type anomaly recorded, yet this was not isolated. Stations WDH and Tianshui (TSE), in Gansu province, were all located to the north of the main epicenter, and station TSE was to the north of station WDH, which was nearby the NWW-striking western Qinling rupture belt and was 452 km from the main epicenter. At station TSE, a rise anomaly beyond 1% appeared during two months period preceding the main shock, which was the most prominent AR change in the last ten years at the station. The phenomenon of mostly drop-type AR anomalies coincided with previous researches, and the spatial distribution of the medium-term anomalies at the four stations well tallied, in the range of - 400 km, with that before EQs with magnitude Ms>7.0.

For change of patterns, a drop-type or rise-type change, of the medium-term AR anomalies which were processed by using the normalized variation rate method (NVRM)[9,7], Du et al.[10,7] got the following statistic results: for Ms ≥ 7.0 EQs, about 100% of the anomalies in the range of 150 km from epicentral areas are negative (a drop-type pattern) and about 71% of the anomalies in the range of 400 km are still negative. The reasons on the change patterns of the anomalies was theoretically explained by papers[10, 7].

4. Large-amplitude anomalies

At station CDU, which was the nearest station to the EQ epicenter, the anomaly amplitude reached up to -5.5%. At stations GAZ and WDH, which were farther away from the EQ epicenter than station CDU, the anomaly amplitudes reached up to -5.3% −-4.9% and 2.9%, respectively. The mean of the anomaly amplitudes was larger than that before the1976 M7.8 Tangshan EQ, and the anomaly amplitudes of the three stations decreased with the increase of epicentral distance. The anomaly recorded at JYO was small in amplitude, only -1.1%, although it was nearby the main EQ epicenter area (according to an investigation after the EQ, the measuring instrument at the station was not in good operation at that time, with a fixed error).

The relationship between AR anomaly amplitude and EQ magnitude has been studied by numerous scholars[5-7]. The anomaly amplitude before the Wenchuan great EQ was in accordance with the previous researches.

5. Anisotropic AR changes

Two observation channels, along N58°E and N49°W directions, are employed at station CDU. The anomaly amplitude recorded through the N58°E channel was -5.5%, whereas no anomaly was recorded through the N49°W channel. According to previous works[2,11], this anisotropic AR changes roughly indicated that the underground media here had been under the action of the maximum compressive stress in the NW-SE direction during the period from about Aug. 2006 to the occurrence of the main shock. At station GAZ, two channels, along N30°E and N60°W directions, are employed. The anomaly amplitude recorded through the N30°E channel was -5.3%, and that recorded through the N60°W channel was - 4.9%. This indicated that the media here had been under the maximum compressive stress in the NWW-SEE direction during the period. The maximum compressive stress directions that were revealed by the AR changes at the two stations basically corresponded to the P-axis azimuth of the main shock[12]. Station JYO was near the northern aftershock area of the great EQ, and two channels, along the N70°W and N10°E directions, are employed at the station. The comparison between anomaly amplitudes recorded through the two channels was still credible as the measuring instrument at the station had just a fixed error. The anomaly amplitude recorded through the N70°W channel was -1.1% and no anomaly was recorded through the N10°E channel. This indicated that the media here had been under the maximum compressive stress close to the NS direction during the period, which well corresponded to the P-axis azimuths of most strong aftershocks in the northern aftershock area.

In summary, the behaviors of these medium-term AR anomalies, such as their spatial distribution within 400 km, their tectonic relevance to the Songpan-Gazi active block, and their amplitude attenuation with increasing distance, as well as their synchronous, mostly descending, large-amplitude and anisotropic changes, strongly support that these anomalies are indeed related to the focal processes of the main shock and strong aftershocks. Such locally concentrative AR anomalies as recorded at the four stations have not been succeeded in identifying in areas beyond 400 km from the EQ epicenter.

2.1.2.2 Short-term AR anomalies before the EQ

During the short-term period before the occurrence date of the great EQ, no obvious anomaly, like these recorded at station CLH (72km) in Hebei province before the 1976 M7.8 Tangshan EQ and at station WDH (110 km) before the 1976 M7.2 Songpan-Pingwu EQs (Fig.2), was recorded at the six stations within 400 km from the EQ epicenter. Upward AR changes commenced at station CDU from March 2008 and at station JYO from April 2008 (Fig.5a-b and Fig.6). The patterns consisting of an initial medium-term drop change followed by a short-term rise change before the EQ are consistent with these of the electrical resistivity change within an EQ focal area that are forecasted by DD model[13-14]. In fact, such patterns appeared often nearby epicentral areas of previous strong EQs[9, 2,7]. However, the rise change, upon which the main shock occurred, did not satisfy the anomaly criterion in amplitude, so no sufficient precursory information could be confidently detected in the period when approaching the EQ.

Du et al[3, 15] studied the spatial distribution characteristics of imminent AR anomalies before the moderate, strong EQs that occurred in the continent of China. As a result, it was got that the spatial distribution of the anomalies was influenced by mainly active faults around EQ focal areas and focal mechanisms. The influences include: (1) the anomalies appear mostly along or nearby the faults; (2) most anomalies are distributed in the two areas that are symmetrical about an epicenter and that azimuthally tie to the P- or T-axis areas that correspond to the focal fault movement; (3) If there is an active fault between a station and an epicentral area, and the fault strike is along or close to the azimuth of the foregoing P- or T-axis, then, no imminent anomaly is usually recorded at the station in the period when approaching the EQ, or an anomaly is generally weak in amplitude.

For example, in the Songpan-Pingwu area in Sichuan Province, three EQs with magnitude of Ms7.2, Ms6.7 and Ms7.2 occurred successively within 8 days in August 1976 at almost the same location (Fig.7), under the action of the compressive stresses in the NE direction (for the previous Ms7.2 EQ) and near to the EW direction (for the following Ms6.7/7.2 EQs)[16].

And the P-axis azimuths of the three EQs were 63°, 101° and 95°, respectively, resulting in an right-lateral movement of the NNW-striking Huya fault (i.e., the focal fault) when the previous Ms7.2 EQ occurred, and then an left-lateral movements of the fault when the following Ms6.7 and Ms7.2 EQs occurred[16]. At that time, three stations were in observation within a range of ~190 km from the three epicenters: station Songpan (SPN, 45km, and it was abandoned shortly afterwards) in Sichuan Province and station Lixian (LXN, 190 km, and abandoned in the 1990s) and station WDH (110 km), in Gansu Province. There were obvious differences between AR changes recorded at station SPN and those recorded at station LXN (Fig. 8). Station SPN was located on the west side of the Minjiang River fault (MRF) with the NNE strike, whereas the epicenters were to the east of the fault. As a result, no imminent AR anomaly was recorded at station SPN when approaching the occurrence date of the first Ms7.2 EQ, though the station was only 45 km from the epicenter, whereas at LXN station, a drop anomaly with amplitude of -0.8% was recorded during 4 days before the EQ in despite of being 190 km away. Contrarily, for the subsequent Ms6.7 and Ms7.2 EQs, a drop anomaly with amplitude of -3.5% was clearly captured at station SPN during 3–4 days before them. This anomaly amplitude was much greater than that recorded at station LXN before the first Ms7.2 EQ. No imminent anomaly appeared at station LXN for the latter EQs. At station WDH, which was located between station LXN and the three epicenters and was near to the north side of the White Dragon River fault (WDRF) with the EW strike, the AR changes recorded before the three events were similar to those seen at station LXN.

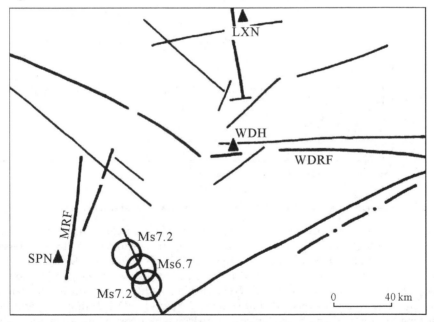

Fig. 7. Distribution of epicenters, active faults and AR stations

According to papers[4, 17], imminent anomalies recorded at underground water level and water chemistry stations before the moderate, strong EQs which occurred in the continent of China also demonstrated such spatial distribution as AR imminent anomalies, which was related to mainly active faults of epicentral areas and EQ focal mechanisms. From loading experiments of rock sample[18], the crust medium nearby active faults are susceptive to stress disturbances during the loading processes, therefore, the AR and underground-water anomalies which is related to the geo-stress changes are easy recorded nearby the faults. Paper [19] calculated the strain distribution in a geologic body model with a fault using the elasto-plastic 2-D finite element method, and then the calculated strain values are converted into relative AR changes. It can be seen from the spatially non-uniform distribution of these calculated AR changes that the spatial distribution of AR changes which are related mainly active faults around epicentral areas and EQ focal mechanisms, as described in papers [3,15,7], can be well explained. In fact, the spatially non-uniform distribution of imminent anomalies in observation of underground water level and water chemistry, as described in papers [4, 17], can be well explained based on the calculated strain distribution by paper [19] also.

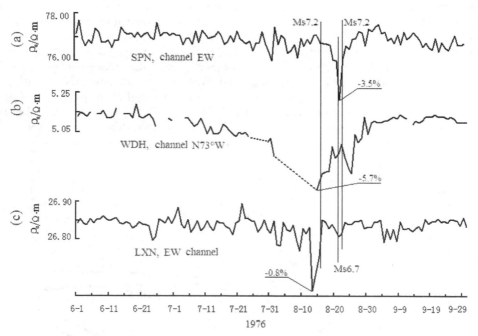

Fig. 8. AR daily-mean curves of SPN (a), WDH (b) and LXN (c) stations

Based on the above-mentioned research works, the lack of imminent AR anomalies before the Wenchuan great EQ can be roughly explained. The main shock and strong aftershocks occurred to the west of the NE-striking Doujiang Weir-An county fault; station GAZ was located on the southwestern side of the NW-striking Xianshui River fault, and station WDH was located on the northern side of the NWW-striking WDRF (Fig.4 and Fig.7). Hence, the reason why no anomaly was recorded at stations GAZ and WDH during the period when approaching the great EQ can be explained. The reason why no anomaly was seen at stations MNI and XCM can be explained also. Stations CDU and JYO were located to the east of the Doujiang-Weir-An county fault, near the main epicenter and in the area liable to record the anomaly related to the main shock, but at the two stations only weak upward changes were recorded during the period. The reason for this remains to be explained and requires additional research.-A possible reason is affected by secondary faults nearby the two stations.

2.2 Medium-term AR anomalies discerned before EQs

On Oct. 25, 2003, two EQs with magnitude of Ms6.1 and Ms5.8 occurred in Minle-Shandan area in Gansu province of China. This year, on Jul. 21, one Ms 6.2 EQ occurred in the Dayao area in Yunnan province, and on Oct. 6, a Ms6.1 EQ occurred also in Dayao area. In fact, the medium-term AR anomalies were well discerned before the two groups of strong EQs, and the EQ locations and magnitudes were successfully forecasted on a one-year time scale, in Nov. of 2002.

Within the range of ~400 km from the epicenters of the Ms 6.1/5.8 Minle- Shandan EQs, there were 6 AR stations, such as Shandan (SHD), Wuwei (WWE), Jayuguan (JYG), Lanzhou (LZL), Linxia (LNX) and Dingxi (DGX) stations, which are all set by China Earthquake Administration (CEA). Of 6 stations, the previous 5 stations, not including DGX station, were in good observation at that time. Of the 5 stations, reliable medium-term AR anomalies, drop-type AR changes, appeared at station SHD during the end of 2002. This station has always kept an observational environment which is up to the technical requirement of seismic geo-electrical station in the long-term observation. Figure 9 shows AR normalized variation rate curves of three channels of this station from Aug. of 1988 to Oct. of 2002, based on monthly AR averages, which are processed by the normalized variation rate method (NVRM)[9,7] in Nov. of 2002. It can be seen from the curves that notable medium-term or short-term AR anomalies which were up to the identification criterion for NVRM anomaly, beyond the threshold value of ± 2.4, appeared before several EQs with magnitude of Ms ≥ 5.0 around the station from 1990 to 2001. In 2002, two drop-type AR anomalies appeared again through channels EW and NW of the station, and a rise-type anomaly did through channel NS. These anomalies were new, following the rise-type anomalies of three channels in the end of 2001 which corresponded to the Nov. 2011 Kunlunshan mountain Ms8.1 EQ far—one type of anomalies which had nothing with the focal process of the great EQ[20, 7-8]. According to the past EQ cases of the station where AR anomalies corresponded to EQs around and the research results in papers [8,10], Du et al. forecasted in Nov. 2002 that one strong EQ will occur nearby the station in 2003 year.

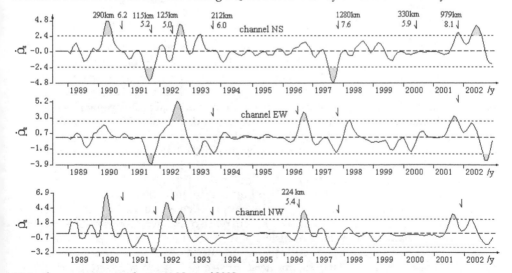

Note: these curves were drown in Nov. of 2002

Fig. 9. NVRM curves of AR changes of SHD station (from Aug. of 1988 to Oct. of 2002)

In Nov. of 2002, Du *et al.* also detected that AR anomalies appeared at Panzhihua (PAH) and Yuanmou (YNM) in Yunnan province. The AR data, monthly mean data, observed by the two stations were processed using NVRM. Figure 10 gives the NVRM curves of PAH station which were reprocessed in 2004. This station was settled in 1970s, around which many EQs with magnitude of Ms ≥ 5.0 have occurred since then, accordingly, at this station AR anomalies have been recorded before the EQs. According to the previous EQ monitoring efficiency of the station, Du *et al.* forecasted in Nov. of 2002 that the AR upward anomaly of 2002 at this station, as seen in figure 10 (the lower curve in the figure), possibly indicated an future strong EQ nearby the station. Besides, at YNM station which was close to the PAH station, AR upward anomalies appeared in the end of 2002 also. Thus, the credibility for the expected EQ was increased.

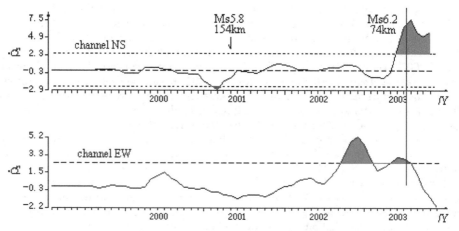

Note: these curves were drown in Nov. of 2004

Fig. 10. NVRM curves of AR changes of PAH station (from Jan. of 199 to Dec. of 2003)

Apart from these above-mentioned, Du *et al.* processed the observation data of other AR stations in the continent of China in Nov. of 2002, as a result, in other places some AR anomalies appeared in 2002 also. Thus, Du *et al.* submitted an EQ forecast view to CEA, a formal written report in Nov. of 2002. In the report, Du *et al.* believed that several EQs probably occurred in 2003 year within 4 regions with the maximal radius being less than 100 km and the minimal radius being less than 60 km, and expected EQ magnitudes were Ms6.0±, Ms≥6.0, Ms5~6 and Ms≥5 (Fig.11), respectively. As a result, the EQs as expected in this report indeed occurred in the former two regions in 2003. It is the case that the Oct. 25, 2003, Ms6.1 and 5.8 Minle~Shandan EQs just occurred in the forecast region with magnitude of Ms6.0±, in Gansu province; and the Jul. 21 Ms6.2 and Oct. 10 Ms6.1, 2003, Dayao EQs just occurred in the forecast region with magnitude of Ms≥6.0, in Yunnan province (Fig.11). In figure 11, the two forecast regions were marked in green color, painted in Nov. 2002, and the two solid circles in red are the locations of occurrence of the two groups of EQs. In other two forecast regions with magnitude of Ms5~6 and Ms≥5, the expected EQs did not occur, and still nearby the forecast region with magnitude of Ms 5~6, in Shanxi province, one Ms5.1 EQ happened in Nov. 2003. The former two strong EQs are successfully predicted, which locations and magnitudes are properly estimated and which occurred in 2003 year, a one-year time scale prediction, therefore, it is reasonable to believe that the AR changes related to the two strong EQs were truly recorded before the occurrence of the two EQs. In 2005, the case for forecasting the two strong EQs was in public reported in paper [21].

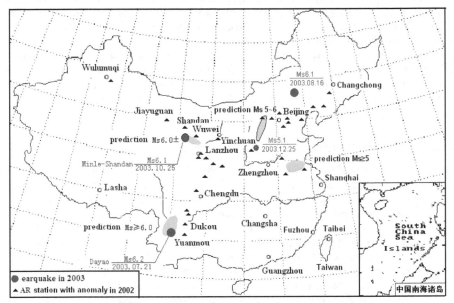

Note: (1) Red solid circles are EQ epicenters that occurred in 2003; (2) Green areas are forecast areas where EQ will occur in 2003; (3) Black solid triangles are AR stations where AR anomaly appeared in 2002

Fig. 11. Distributions of epicenters, AR stations and forecast areas

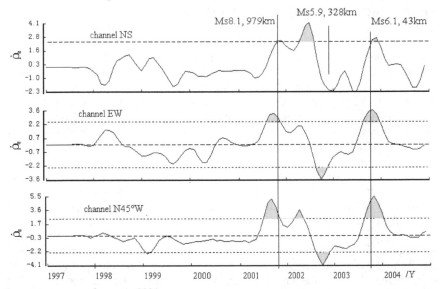

Note: these curves were drown in 2004

Fig. 12. NVRM curves of AR changes of SHD station (from 1997 to 2004)

After the Ms6.1 and 5.8 Shandan-Minle EQs, Du *et al* again processed the AR data observed at SHD station (43 km) from the two epicentral areas 1997 to 2004 using NVRM, as a result, it can be seen from figure 12 that the medium-term drop-type to short-term rise-type AR changes are similar to the AR changes in the focal area as foretold by the DD model (Dilatancy-Diffusion Model)[13-14] in appearance, a pattern consisting of an initial medium-term fall followed by a short-term rise.

People may ask why the May 12, 2008, Ms8.0 EQ was not forecasted in the medium-term period before the great EQ, which just occurred around/nearby 6 AR stations in this area? In fact, authors had no time to process and analyze the AR data observed at the 6 AR stations in those days, and the AR changes of the stations were analyzed and studied only after the great EQ[7].

3. Anisotropic AR changes related to strong EQ

Kraev studied the calculating method of AR in a homogeneous anisotropy medium[22]. Later, Brace *et al.*[23], Chen *et al.*[24] and Lu *et al.*[25] reported that during the loading process of most rock (soil) samples, AR changes showed the directional changes which were associated with the maximum loading direction. In recent years, people try to explore related-earthquake anisotropic AR changes, which would have an important significance to understand the stress status in/nearby the EQ focal region, to explain the reason of related-earthquake AR changes and to forecast an EQ. However, it is very difficult to detect the anisotropic changes from actual EQ cases. Such researches were reported only in China because China has long made a lot of fixed-site AR observations. Qian *et al.*[26] and Du *et al.*[9, 11] reported that the anisotropic AR changes that were associated with the maximum compressional stress (P-axis) azimuths of EQ focal mechanism solutions, and Du *et al.*[2, 27] try to explain the anisotropic AR changes in theory.

3.1 Analysis of EQ cases on anisotropic AR changes

The studied EQ cases are picked out according to the following three principles: ①strong EQs with magnitude of Ms ≥ 5.5, ②EQs nearby AR stations, and ③AR changes in the later stages of EQ preparation. From the formula to estimate the focal body radius L[29], $Ms=3.3+2.1 \, Log \, L$, for Ms5.5 EQ L is no more than 15 km, for Ms 6~7 EQ L is no more than 60 km and for Ms7.8 EQ L is about 140 km. According to Du *et al.*[7, 10], AR anomalies within tens km for Ms 6~6.9 EQs and within 150 km for Ms ≥ 7.0 EQs are mostly characterized by drop-type changes in the medium-term stages of EQ preparation. The underground medium commonly contains rich water, so the concentrative range of the drop-type AR changes is well coincident with the focal body radius L. This indicates how we pick out the EQs nearby stations.

It is very important in study of anisotropic AR changes to distinguish the precursory AR anomaly related to strong EQs from observational AR data. Therefore, NVRM is usually used to process data. Du *et al.*[11] have studied anisotropic AR changes, using NVRM, recorded at 41 stations before 27 Ms ≥ 5.5 EQs that occurred in the Chinese mainland. The results show that for over 95% of the stations, the anisotropic AR changes appeared during the later stages of EQ preparation, which were obviously related to the P-axis azimuth of EQ focal mechanism solution. The behaviors of these anisotropic changes are that the AR change recorded through the channel perpendicular (or almost perpendicular) to the azimuth is a maximum in amplitude, whereas that recorded through the channel along (or close to) the azimuth is a minimum.

For example, the two observation channels, N20°E and N70°W channels, are installed at station PGU that had epicentral distances of 111 and 140 km for the 1976 Ms7.8 Tangshan EQ and Ms7.1 Luanxian aftershock; the P-axis azimuths of the two events were 75° and 297°, respectively, roughly in the EW direction. As a result, the medium-term drop-type and short-term rise-type AR changes recorded through N20°E channel (with a near NS direction) prior to the EQs were greater in amplitude than those through N70°W channel (with a near EW direction) (Fig.13). As another example, three channels, NS, EW and N45°W channels, are installed at station SHD that had an epicentral distance of 43 km for the 2003 Ms6.1 Minle- Shandan EQ; the P-axis azimuth of this EQ was 65°. As a result, the medium-term drop-type and short-term rise-type AR changes recorded through N45°W channel prior to the EQ were greater in amplitude than those through EW channel (Fig. 12).

It is obvious that the relationship between the anisotropic AR changes and the P-axis azimuth from actual EQ cases agrees well with the relationship between the directional AR changes and the maximum loading direction in most experiments of water-bearing rock (or soil) samples. This proves that such AR changes are just related to the EQ preparation process.

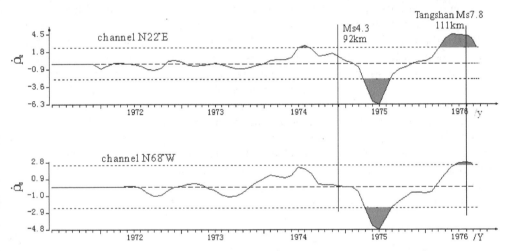

Fig. 13. NVRM curves of AR changes of station PGU for the 1976 Ms 7.8 Tangshan EQ

3.2 Theoretical analysis on anisotropic AR changes

According to DD model, the micro cracks inside the underground medium fast and nonlinearly develop immediately before the main rupture within an EQ focal area, their strikes align predominately in a certain direction and underground water fast come in them. Barsukov[29] interpreted a larger amplitude change in resistivity on the assumption that the micro crocks are tortuously linked each other, and underground water comes in them, as a result, the conductive aisles inside the medium are formed. Mei et al.[30] deduced that in the later preparation stages of strong EQs a number of micro-cracks inside the medium of an EQ focal region is increased sharply. The fact that in the medium-term statges before moderate, strong EQs of the Chinese mainland, for larger-magnitude EQs the AR anomaly amplitude increases fast, whereas the anomaly duration increases more slowly[10] supports Mei's deduction. According to Crampin et at.[31], the maximum compressional stress inside the near-surface crust is commonly horizontal, and because of hydrostatic pressure of rock the horizontal cracks are closed and the erect cracks are developed, with the normal of the erect cracks being usually mostly horizontal and the strikes of the cracks being predominantly along in the direction of the maximum compressional stress axis. As a result, this forms an EDA medium with conductive fluid. According to the above-mentioned discussions, we presume that in the later preparation stages of strong EQs there probably exist two physical behaviors inside the medium in and nearby the focal area: (a) the erect micro-cracks is well developed, their number is non-linearly increased, and their strikes are predominantly along the direction of the maximum compressional stress axis. (b) in above-mentioned processes, the macro-cracks are linked each other and the underground water with low resistance comes fast in them. This forms the conductive aisles in the medium. As the result of its sensitivity to water, the electrical resistivity of the medium, therefore, undergoes significant changes.

Based on the two physical behaviors, we approximately regard the medium in/nearby the EQ focal region as a homogeneous azimuthal anisotropic medium. We establish a Descartes coordinate system $o - x_1 x_2 x_3$ on the ground where x_1, x_2 and x_3 are three electrical principal axes, respectively, which x_2 is along the horizontal loading direction (this means the P-axis direction of EQ focal mechanism solution in study of EQ cases), x_1 is horizontal and perpendicular to the direction and x_3 is downwards vertical to the ground surface. ρ_1, ρ_2 and ρ_3 are true resistivity (TR) along the three axes, respectively, and $\rho_1 > \rho_2 = \rho_3$. φ is the angle between ground observation channel X and the axis x_1. Another ground observation channel Y is perpendicular to channel X. Based on the calculating formula for the relative AR changes ($\Delta\rho_s/\rho_s$) in the anisotropic medium by Du et al.[2] (after referring to paper [22]), we can get the relation of relative AR changes ($\Delta\rho_{sx}/\rho_{sx}$ and $\Delta\rho_{sy}/\rho_{sy}$) of observational channels X and Y to relative TR changes ($\Delta\rho_1/\rho_1$ and $\Delta\rho_2/\rho_2$) in two horizontal principal axes directions. When $\varphi = 0°$ (here, channel X is perpendicular to the loading direction and channel Y is along the direction) the relation is as follows:

In x_1 direction (perpendicular to the maximum loading direction)

$$\frac{\Delta\rho_{sx}}{\rho_{sx}} = \frac{\Delta\rho_2}{\rho_2} \qquad \qquad (1)$$

In x_2 direction (along the loading direction)

$$\frac{\Delta\rho_{sy}}{\rho_{sy}} = \frac{1}{2}\left(\frac{\Delta\rho_1}{\rho_1} + \frac{\Delta\rho_2}{\rho_2}\right). \qquad (2)$$

According to most loading experiments of rock (soil) samples, following two inequalities are generally true:

$$\text{When } \frac{\Delta\rho_s}{\rho_s} < 0, \qquad \frac{\Delta\rho_{sx}}{\rho_{sx}} < \frac{\Delta\rho_{sy}}{\rho_{sy}}; \qquad (3)$$

$$\text{When } \frac{\Delta\rho_s}{\rho_s} > 0 \qquad \frac{\Delta\rho_{sx}}{\rho_{sx}} > \frac{\Delta\rho_{sy}}{\rho_{sy}}. \qquad (4)$$

From formulas (1) ~ (3), we can get that the TR changes along the two principal axes satisfy the following inequality:

$$\frac{\Delta\rho_2}{\rho_2} < \frac{\Delta\rho_1}{\rho_1}. \tag{5}$$

Let the AR variation rate $\dot{\rho}_s = \Delta\rho_s\big/\Delta t$ (Δt is a time interval). When the underground medium change from homogeneous ($\rho_{sx} = \rho_{sy}$) into anisotropic ($\rho_{sx} \neq \rho_{sy}$), we can obtain the inequality of $\dot{\rho}_{sx} < \dot{\rho}_{sy}$ from inequality (3), in which $\dot{\rho}_{sx}$ is the AR variation rate of channel X and $\dot{\rho}_{sy}$ is that of channel Y. From the inequality and formulae (1) and (2) as well as the calculating formula for ρ_s in the anisotropic medium, the following inequality is obtained[11]:

$$\dot{\rho}_2 < \frac{\lambda^2}{2\lambda - 1}\dot{\rho}_1. \tag{6}$$

Where $\dot{\rho}_1$ and $\dot{\rho}_2$ are TR variation rates along two electrical principle axes, and a true anisotropic coefficient $\lambda = \sqrt{\rho_2\big/\rho_1}$ ($0 < \lambda < 1$). In the case of $0.5 < \lambda < 1.0$ (viz., $1 < \rho_1/\rho_2 < 4$), we get the following inequality:

$$\dot{\rho}_2 < \dot{\rho}_1. \tag{7}$$

If the AR changes along the loading direction and perpendicular to the direction are all increased the sigh ">"will replace "<" in inequalities (5) and (7), from inequality (4). According to inequalities (5) and (7) and their derivation conditions, we can have the following understandings: anisotropic TR changes in which the change along the loading direction is more prominent than that perpendicular to the direction cause anisotropic AR changes in which the change perpendicular to the direction is more prominent than that in the direction. It is obviously that there are a difference of $90°$ between the directions of both the most prominent AR change (perpendicular to the loading direction) and the most prominent TR change (along the loading direction). This result distinctly explained the relationship between the most loading direction and the anisotropic AR changes when approaching the main rupture of most rock (or soil) samples and theoretically supported the research results from actual EQ cases.

3.3 A Reason for anisotropic AR changes

From the TR relationship among three principal axes in the anisotropic medium, $\rho_1 > \rho_2 = \rho_3$, we assume the shape of a single micro crack as a schistose ellipsoid that takes ρ_1 axis as its rotation axis, and whose radius is a in x_2 direction, b in x_3 direction and c in x_1 direction and a=b>c (an erect crack). And then, we assume that the resistivity in saturated-water cracks (water is rich within the underground medium) is ρ_f, that of framework is ρ_0 and the crack ratio is v. Using Kraev's result[22] we have the approximate formulae of $\Delta\rho/\rho$ versus $\Delta v/v$:

$$\left.\begin{array}{l}\dfrac{\Delta\rho_1}{\rho_1} \approx \dfrac{\rho_f - \rho_\circ}{\rho_\circ/v + \rho_f} \cdot \dfrac{\Delta v}{v} \\[3mm] \dfrac{\Delta\rho_2}{\rho_2} \approx \dfrac{\rho_f - \rho_\circ}{\rho_f/v + \rho_\circ} \cdot \dfrac{\Delta v}{v}\end{array}\right\}. \tag{8}$$

In generally, $\rho_\circ \gg \rho_f$. From formulae (8), $\Delta\rho_1/\rho_1 < 0$ and $\Delta\rho_2/\rho_2 < 0$ when $\Delta v/v > 0$, and they will decrease with $\Delta v/v$ going bigger, which could be why most AR anomalies in/nearby the epicenter region in the late preparation stages of strong EQs are commonly a drop-type pattern; And $\Delta\rho_1/\rho_1 > 0$ and $\Delta\rho_2/\rho_2 > 0$ when $\Delta v/v < 0$, and they increase with $\Delta v/v$ decreasing. This is coincident with the physical analysis. Because $v \ll 1$, in formulae

(8) $\left|\dfrac{\rho_f - \rho_\circ}{\rho_\circ/v + \rho_f}\right| < \left|\dfrac{\rho_f - \rho_\circ}{\rho_f/v + \rho_\circ}\right|$, so for $\Delta v/v > 0$ and $\Delta v/v < 0$ cases $\dfrac{\Delta\rho_2}{\rho_2}\left/\dfrac{\Delta\rho_1}{\rho_1}\right. > 1$ all the time

when the AR changes of channels X and Y, $\Delta\rho_{sx}/\rho_{sx}$ and $\Delta\rho_{sy}/\rho_{sy}$, are all increased or decreased. This is coincident with the physical meaning of inequality (5). Let $\dot\rho_i = \Delta\rho_i/\Delta t$

($i = 1, 2$), the equation, $\dfrac{\dot\rho_2}{\dot\rho_1} \approx \dfrac{\rho_\circ\rho_f}{\left(\rho_\circ v + \rho_f\right)^2}$, can be deduced. Thus we get that $\dot\rho_2/\dot\rho_1 > 1$ all the

time. This is coincident with the physical meaning of inequality (7).

In general, it can be seen that anisotropic TR changes is obviously associated with ρ_f, $\rho_\circ$ and v, which is clear in theory, and anisotropic AR changes arise from anisotropic TR changes. Therefore, the reason for AR changes and their anisotropic changes as well as their pattern (drop-type or rise-type) are clear in theory also.

4. Conclusions

1. The AR changes related to the late preparation process of strong EQs are indeed recorded in china. The two proofs are as follows: (a) Reappearing AR changes are observed before two great EQs. (b) Two strong EQs are successfully predicted using AR changes observed at stations nearby on an one-year time scale, which three elements, such as locations, magnitudes and year 2003, are all right.

2. Of 41 stations in or nearby the epicentral areas of 27 strong EQs, for over 95% stations the anisotropic AR changes are related to the maximum compressional stress directions of the EQ focal mechanism solutions. Their behaviors are: the most prominent AR change appears perpendicular or nearly perpendicular to the direction. The relationship between the anisotropic changes and the direction is well coincident with the directional AR changes in the loading process of most rock (soil) samples. And the relationship can be explained in theory. Therefore, we can confirm that the anisotropic AR changes which are directly associated with the later preparation processes of strong EQs are truly recorded nearby epicentral regions.

3. The reasons for AR changes and their anisotropic changes as well as their pattern (drop-type or rise-type) are clear. The compressional action along the maximum compressional stress direction plays an important role in/nearby the EQ focal region in the later preparation stages of strong EQs. This caused that the micro cracks in the underground medium develop fast in number, and their strikes are predominant along the direction, as a result, conductive aisles in the medium are linked each other and

underground water comes fast in. The physical processes induce the TR changes which drop-type pattern occurs in a medium dilatancy stage and rise-type pattern does in a closure stage of micro cracks, and also induce the anisotropic TR changes in which the most prominent change appears along the maximum compressional stress direction.

4. In a homogeneous anisotropic medium, the AR change is in agreement with the TR change in drop-type or rise-type pattern, hence, a drop-type pattern of AR change appears in a medium dilatancy stage and a rise-type pattern appears in a closure stage of micro cracks. Because of the directional discrepancy with $90°$ angle between the most prominent TR change and the most prominent AR change, the anisotropic AR changes, in which the most prominent change appears perpendicular or nearly perpendicular to the maximum loading direction, just appear.

5. References

[1] Gui X T, Guan H P, Dai J A. The short-term and immediate anomalous pattern recurrences of the apparent resistivity before the Tangshan and Songpan earthquake of 1976. Northwestern Seismological Journal (in Chinese, with an English abstract), 1989, 11(4): 71~75

[2] Du X B, Li N, Ye Q, et al. A possible reason for the anisotropic changes in apparent resistivity near the focal region of strong earthquake. Chinese J Geophys (in Chinese), 2007, 50(6): 1802~1810. http://www.agu.org/wps /cjg, Chinese J Geophys (in English), 2007, 50(6): 1555~1565

[3] Du X B, Zhao H Y, Chen B Z, et al. On the relation of the imminent sudden change in earth resistivity to the active fault and generating-earthquake stress field. Acta Seismolgic Sinica, 1993, 6: 663~673

[4] Du X B, Liu Y W, Ni M K. On the spatial characteristic of the short-term and imminent anomalies of underground water behaviors before strong earthquake. Acta Seismologic Sinica, 1997, 10: 523~533

[5] Qian F Y, Zhao Y L, Yu M M. Anomalous changes in geoelectric resistivity before earthquakes (in Chinese). Sci China Ser B, 1982, 12: 831~839

[6] Qian J D, Chen Y F, Jin A Z. The Application of Geoelectrical Resistivity Method in Earthquake Prediction (in Chinese). Beijing: Seismological Press, 1985. 48~132, 226~266

[7] Du X B. Two types of changes in apparent resistivity in earthquake prediction. Science in China-Series D, 2011, 54(1): 145~156 / doi: 10.1007/s11430-010-4031-y

[8] Du X B, Ren G J, Xue S Z. Study on many kinds of precursory anomalies and trial prediction of strong earthquakes in the continent of China (in Chinese, with an English abstract). Northwestern Seismol J, 1999, 21: 113~122

[9] Du X B, Ruan A G, Fan S H, et al. Anisotropy of the apparent resistivity variation rate near the epicentral region for strong earthquake. Acta Seismologic Sinica, 2001, 14(3): 303~314

[10] Du X B, Xue S Z, Hao Z, et al. On the relation of moderate-short term anomaly of earth resistivity to earthquake. Acta Seismologic Sinica, 2000, 13: 393~403

[11] Du X B, Ma Z H, Ye Q, et al. Anisotropic changes in earth resistivity associated with strong earthquakes. Progress in Geophysics (in Chinese, with an English abstract), 2006, 21(1): 93~100.

[12] Lu H F, Jia D, Wang L S, et al. On the triggering mechanics of Wenchuan earthquake (in Chinese). Geol J Chin Univ, 2008, 14: 133~138

[13] Nur A. Dilatancy, pore fluids, and premonitory variations of ts/tp travel times. Bull Seismol Soc Amer, 1972, 62: 1217~1222

[14] Scholz C H, Sykes L R, Aggarwal Y P. Earthquake prediction: A physical basis. Science, 1973, 181: 803~810

[15] Du X B, Yan Z D, Zou M W, et al. Process of source dynamics of the Jingtai earthquake (M=6.2). Acta Seismologic Sinica, 1994, 7: 379~388

[16] Seismological Bureau of Sichuan Province. The Songpan Earthquake in 1976 (in Chinese). Beijing: Seismological Press, 1979. 4~5, 86~91, 103

[17] Du X B, Zhang X J, Zhang H, et al. The spatial characteristics of the short-term and imminent anomalies of water radon before earthquake in the mainland of China. Acta Seismologic Sinica, 1996, 9: 461~470

[18] Ma J, Ma S L, Liu L Q. The stages of anomalies before an earthquake and the characteristics of their spatial distribution (in Chinese). Seismol Geol, 1995, 17: 363~371

[19] Zheng G. L., Du X. B., Chen J. Y., et al. Influence of active faults on erathquake-related anomalies of geo-electric resistivity (in Chinese, with an English abstract). Acta Seismol Sinic, 2011, 33: 187~197

[20] Du X B, Tan D C. On the temporal and spatial clusters of one-year scale anomalies of earth-resistivity and the relation to seismicity (in Chinese, with an English abstract). Earthquake Research in China, 2000, 6: 283~292

[21] Ye Q, Du X B, Chen J Y, et al. One-Year Prediction for the Dayao and Minle-Shandan Earthquakes in 2003 (in Chinese, with an English abstract). Journal of seismological research, 2005, 28(13): 226~230

[22] A.P.Kraev., 1951.Geoelectrics Principle. Moscow State's technological and theoretical liber press of Soviet Union, 10~50

[23] W.F. Brace., 1968. Electrical resistivity changes in saturated rocks during fracture and fractional sliding, J. Geophys. Res., Vol. 73, 1433~1444

[24] Chen D Y, Chen F, Wang L H, 1983. Study of rock resistivity under uniaxial press-Anisotropy of resistivity. Acta Geophysica Sinica, Vol. 26 (Supp.): 783~792

[25] Lu Y Q, Qian J D, Liu J Y. An experimental study on the precursory features of apparent resistivity and acoustic emission of large scale of granite specimen during the process of slowly dilatancy rupturing (in Chinese, with an English abstract). Northwestern Seismol J, 1990, 12: 35~41

[26] Qian F Y, Zhao Y L, Huang Y N. Anisotropic parameters calculation of earth resistivity and seismic precursory examples · Acta Seismologica Sinica, 1996, 9(4): 617~627

[27] Du X B, Ye Q, Ma Z H, et al. The detection depth of symmetric four-electrode resistivity observation in/near the epicentral region of strong earthquakes (in Chinese, with an English abstract). Chin J Geophys, 2008, 51: 1943–1949. http://www.agu.org/wps /cjg, Chinese J Geophys (in English), 2008, 51(6): 1220~1228

[28] Guo Z J, Qin F Y. 1979. Physics of Earthquake Source (in Chinese). Beijing: Seismological Press, 100~170

[29] O.M. Barsukov, 1979.A possible cause of electrical precursors to earthquake. Earth's Physics, Vol. 8, 85~90

[30] Mei S R, Fen D Y, Zhang G M, 1993. Introduction of Earthquake Research in China. Beijing: Seismological press, 302~307

[31] Crampin S, Evan R, Atkins B.K., 1984. Earthquake prediction: a new physical basis. Geophys. J.R.astr.soc, Vol.76, 147~156

Permissions

The contributors of this book come from diverse backgrounds, making this book a truly international effort. This book will bring forth new frontiers with its revolutionizing research information and detailed analysis of the nascent developments around the world.

We would like to thank Sebastiano D'Amico, for lending his expertise to make the book truly unique. He has played a crucial role in the development of this book. Without his invaluable contribution this book wouldn't have been possible. He has made vital efforts to compile up to date information on the varied aspects of this subject to make this book a valuable addition to the collection of many professionals and students.

This book was conceptualized with the vision of imparting up-to-date information and advanced data in this field. To ensure the same, a matchless editorial board was set up. Every individual on the board went through rigorous rounds of assessment to prove their worth. After which they invested a large part of their time researching and compiling the most relevant data for our readers. Conferences and sessions were held from time to time between the editorial board and the contributing authors to present the data in the most comprehensible form. The editorial team has worked tirelessly to provide valuable and valid information to help people across the globe.

Every chapter published in this book has been scrutinized by our experts. Their significance has been extensively debated. The topics covered herein carry significant findings which will fuel the growth of the discipline. They may even be implemented as practical applications or may be referred to as a beginning point for another development. Chapters in this book were first published by InTech; hereby published with permission under the Creative Commons Attribution License or equivalent.

The editorial board has been involved in producing this book since its inception. They have spent rigorous hours researching and exploring the diverse topics which have resulted in the successful publishing of this book. They have passed on their knowledge of decades through this book. To expedite this challenging task, the publisher supported the team at every step. A small team of assistant editors was also appointed to further simplify the editing procedure and attain best results for the readers.

Our editorial team has been hand-picked from every corner of the world. Their multi-ethnicity adds dynamic inputs to the discussions which result in innovative outcomes. These outcomes are then further discussed with the researchers and contributors who give their valuable feedback and opinion regarding the same. The feedback is then collaborated with the researches and they are edited in a comprehensive manner to aid the understanding of the subject.

Apart from the editorial board, the designing team has also invested a significant amount of their time in understanding the subject and creating the most relevant covers. They scrutinized every image to scout for the most suitable representation of the subject and create an appropriate cover for the book.

The publishing team has been involved in this book since its early stages. They were actively engaged in every process, be it collecting the data, connecting with the contributors or procuring relevant information. The team has been an ardent support to the editorial, designing and production team. Their endless efforts to recruit the best for this project, has resulted in the accomplishment of this book. They are a veteran in the field of academics and their pool of knowledge is as vast as their experience in printing. Their expertise and guidance has proved useful at every step. Their uncompromising quality standards have made this book an exceptional effort. Their encouragement from time to time has been an inspiration for everyone.

The publisher and the editorial board hope that this book will prove to be a valuable piece of knowledge for researchers, students, practitioners and scholars across the globe.

List of Contributors

Tomohiro Hasumi
Division of Environment, Natural Resources and Energy, Mizuho Information and Research Institute, Inc., Tokyo, Japan

Chien-chih Chen
Department of Earth Sciences and Graduate Institute of Geophysics, National Central University, Jhongli, Taoyuan, Taiwan

Takuma Akimoto
Department of Mechanical Engineering, Keio University, Yokohama, Japan

Yoji Aizawa
Department of Applied Physics, Advanced School of Science and Engineering, Waseda University, Tokyo, Japan

Ayten Yiğiter
Hacettepe University, Faculty of Science, Department of Statistics Beytepe-Ankara, Turkey

I.N. Tikhonov
Institute of Marine Geology and Geophysics FEB RAS, Yuzhno-Sakhalinsk, Russia

M.V. Rodkin
Institute of Marine Geology and Geophysics FEB RAS, Yuzhno-Sakhalinsk, Russia
International Institute of Earthquake Prediction Theory and Mathematical, Geophysics RAS, Moscow, Russia

Dumitru Stanica and Dragos Armand Stanica
Institute of Geodynamics of the Romanian Academy, Romania

R. G. M. Crockett
University of Northampton, United Kingdom

Elvis Pontes, Anderson A. A. Silva, Adilson E. Guelfi and Sérgio T. Kofuji
Laboratory of Integrated Systems, Polytechnic School of the University of São Paulo, Brazil

Ming-Ching T. Kuo
National Cheng Kung University, Taiwan

Giuseppina Immè and Daniela Morelli
Dipartimento di Fisica e Astronomia Università di Catania - INFN Sezione di Catania, Italy

Asta Gregorič, Boris Zmazek, Sašo Džeroski, Drago Torkar and Janja Vaupotič
Jožef Stefan Institute, Ljubljana, Slovenia

Konstantinos Eftaxias
University of Athens, Faculty of Physics, Department of Solid State Section, Panepistimiopolis Zofrafos, Athens, Greece

Du Xuebin, An Zhanghui, Fan Yingying, Liu Jun, Chen Junying and Tan Dacheng
Lanzhou Base of Institute of Earthquake Science, CEA, Lanzhou, China

Ye Qing
China Earthquake Networks Center, China Earthquake Administration, Beijing, China

Yan Rui
Earthquake Administration of Beijing Municipality, Beijing, China

Printed in the USA
CPSIA information can be obtained
at www.ICGtesting.com
JSHW011437221024
72173JS00004B/844